Progress in Inflammation Research

Series Editor

Prof. Michael J. Parnham PhD
Senior Scientific Advisor
GSK Research Centre Zagreb Ltd.
Prilaz baruna Filipovića 29
HR-10000 Zagreb
Croatia

Advisory Board

G. Z. Feuerstein (Wyeth Research, Collegeville, PA, USA)
M. Pairet (Boehringer Ingelheim Pharma KG, Biberach a. d. Riss, Germany)
W. van Eden (Universiteit Utrecht, Utrecht, The Netherlands)

Forthcoming titles:

Chemokine Biology: Basic Research and Clinical Application, Volume II: Pathophysiology of Chemokines, K. Neote, G. L. Letts, B. Moser (Editors), 2006
The Resolution of Inflammation, A. G. Rossi, D. A. Sawatzky (Editors), 2006

(Already published titles see last page.)

In Vivo Models of Inflammation

2nd Edition, Volume II

Christopher S. Stevenson
Lisa A. Marshall
Douglas W. Morgan

Editors

Birkhäuser Verlag
Basel · Boston · Berlin

Editors

Christopher S. Stevenson
Novartis Institutes for Biomedical Research
Respiratory Disease Area
Novartis Horsham Research Centre
Wimblehurst Road
Horsham, West Sussex
United Kingdom

Lisa A. Marshall
Johnson and Johnson PRD
Welsh & McKean Rd
Spring House, PA 19477
USA

Douglas W. Morgan
Portfolio, Program and Alliance
Management
BiogenIdec
14 Cambridge Center
Cambridge, MA 02142
USA

A CIP catalogue record for this book is available from the Library of Congress, Washington D.C., USA

Bibliographic information published by Die Deutsche Bibliothek
Die Deutsche Bibliothek lists this publication in the Deutsche Nationalbibliografie;
detailed bibliographic data is available in the internet at http://dnb.ddb.de

ISBN-10: 3-7643-7757-7 Birkhäuser Verlag, Basel – Boston – Berlin
ISBN-13: 978-3-7643-7757-1

© 2006 Birkhäuser Verlag, P.O. Box 133, CH-4010 Basel, Switzerland
Part of Springer Science+Business Media
Printed on acid-free paper produced from chlorine-free pulp. TCF ∞
Cover design: Markus Etterich, Basel
Cover illustration: with friendly permission of Christopher S. Stevenson
Printed in Germany
ISBN-10: 3-7643-7757-7
ISBN-13: 978-3-7643-7757-1

e-ISBN-10: 3-7643-7760-7
e-ISBN-13: 978-3-7643-7760-1

9 8 7 6 5 4 3 2 1

www.birkhauser.ch

Contents

Volume II

(contents of volume I see last page)

List of contributors

Lawrence S. Chan, Department of Dermatology, University of Illinois at Chicago, College of Medicine, 808 South Wood Street, MC624, Chicago, IL 60612, USA; e-mail: larrycha@uic.edu

Azzeddine Dakhama, Division of Cell Biology, Department of Pediatrics, National Jewish Medical and Research Center, 1400 Jackson Street, Denver, Colorado 80206, USA; e-mail: dakhamaa@njc.org

James L. Ellis, Surface Logix Inc, 50 Soldiers Field Place, Brighton, MA 02139, USA; e-mail: jellis@surfacelogix.com

Erwin W. Gelfand, Division of Cell Biology, Department of Pediatrics, National Jewish Medical and Research Center, 1400 Jackson Street, Denver, Colorado 80206, USA; e-mail: gelfande@njc.org

Pierangelo Geppetti, Clinical Pharmacology Unit, Department of Critical Care Medicine and Surgery, University of Florence, Viale Pieraccini 6, 50139 Florence, Italy; e-mail: gpp@ifeuniv.unife.it

H. Andreas Kalmes, Department of Inflammation Research, Amgen Washington Inc., Seattle, Washington, 98119, USA; e-mail: kalmesa@amgen.com

Serena Materazzi, Clinical Pharmacology Unit, Department of Critical Care Medicine and Surgery, University of Florence, 50139 Florence, Italy; e-mail: serena.materazzi@unifi.it

Sreekant Murthy, Division of Gastroenterology and Hepatology and Office of Research, Drexel University College of Medicine, Philadelphia, USA; e-mail: sm53@drexel.edu

Paola Nicoletti, Clinical Pharmacology Unit, Department of Critical Care Medicine and Surgery, University of Florence, 50139 Florence, Italy; e-mail: paola.nicoletti@unifi.it

William M. Selig, NitroMed Inc, 125 Spring Street, Lexington, MA 02421, USA; e-mail: wselig@nitromed.com

Christopher S. Stevenson, Novartis Institutes for Biomedical Research, Respiratory Disease Area, Novartis Horsham Research Centre, Wimblehurst Road, Horsham, West Sussex, RH12 5AB, United Kingdom; e-mail: christopher.stevenson@novartis.com

Christopher F. Toombs, Ikaria, Inc., 1616 Eastlake Avenue East, Suite 340, Seattle, Washington, 98102, USA; e-mail: christopher.toombs@ikaria.com

Marcello Trevisani, Center of Excellence for the Study of Inflammation, University of Ferrara, Ferrara, Italy; e-mail: tvm@unife.it

David C. Underwood, Center for Excellence in Drug Discovery, Respiratory Diseases, GlaxoSmithKline Pharmaceuticals, 709 Swedeland Road, King of Prussia, PA 19406, USA

Eric T. Whalley, BiogenIdec, 14 Cambridge Center, Cambridge, MA 02142, USA; e-mail: eric.whalley@biogenidec.com

Preface to the first edition

The purpose of this volume in the series *Progress in Inflammation Research* is to provide the biomedical researcher with a description of the state of the art of the development and use of animal models of diseases with components of inflammation. Particularly highlighted are those models which can serve as *in vivo* correlates of diseases most commonly targeted for therapeutic intervention. The format is designed with the laboratory in mind; thus it provides detailed descriptions of the methodologies and uses of the most significant models. Also, new approaches to the development of future models in selected therapeutic areas have been highlighted. While emphasis is on the newest models, new information broadening our understanding of several well-known models of proven clinical utility is included. In addition, we have provided coverage of transgenic and gene transfer technologies which will undoubtedly serve as tools for many future approaches. Provocative comments on the cutting edge and future directions are meant to stimulate new thinking. Of course, it is important to recognize that the experimental use of animals for human benefit carries with it a solemn responsibility for the welfare of these animals. The reader is referred to the section on current regulations governing animal use which addresses this concern.

To fulfill our purpose, the content is organized according to therapeutic areas with the associated models arranged in subcategories of each therapeutic area. Concepts presented are discussed in the context of their current practice, including intended purpose, methodology, data and limitations. In this way, emphasis is placed on the usefulness of the models and how they work. Data on activities of key reference compounds and/or standards using graphs, tables and figures to illustrate the function of the model are included. The discussions include ideas on a given model's clinical correlate. For example, we asked our contributors to answer this question: How does the model mimic what is found in human clinical practice? They have answered this question in many interesting ways.

We hope the reader will find the information presented here useful for his or her own endeavours investigating processes of inflammation and developing therapeutics to treat inflammatory diseases.

October, 1998
Douglas W. Morgan
Lisa A. Marshall

Preface to the second edition

Since our first edition of "*In Vivo* Models of Inflammation" published in 1999, there has been amazing progress, and an abundance of exciting new information in inflammation research: new technologies, new therapeutics, new understanding of inflammatory processes, ... and on and on, have emerged in the past 6 years. Supporting all of this are the fundamentals of inflammation research, i.e., the animal models, known mechanisms, and therapeutic standards, that have continued to provide the basis for generating these advances. Given the great progress, we have chosen to provide a second edition to our original text.

The second edition of "*In Vivo* Models of Inflammation" comes to you in two volumes and provides an update of the models included in first edition with expanded coverage and more models. Again, these volumes emphasize the standard models regarded as the most relevant for their disease area. The intent is to provide the scientist with an up-to-date reference manual for selecting the best animal model for their specific question. Updates on previously described models are specifically focused on references to any additional pharmacology that has been conducted using these systems. The sections on arthritis models have been expanded and now include models relating to osteoarthritis. New areas described herein include models of neurogenic, cancer, and vascular inflammation. Additionally, coverage of *in vivo* technologies includes updates on transgenic and gene transfer technologies, and has also been expanded to include chapters on stem cells and nanotechnologies.

The second edition continues to emphasize that conducting *in vivo* research carries with it a great responsibility for animal respect and welfare. The coverage of this concern has been extended to include chapters describing current regulations in the United States, the United Kingdom, and Japan.

The ultimate aim of the second edition is to provide current best practices for obtaining the maximum information from *in vivo* experimentation, while preserving the dignity and comfort of the animal.

We hope the information provided here helps in advancing the reader's endeavors in investigating processes of inflammation and in developing therapeutics to treat inflammatory diseases.

May, 2006

Christopher S. Stevenson
Lisa A. Marshall
Douglas W. Morgan

Asthma

William M. Selig[1], Eric T. Whalley[2] and James L. Ellis[3]

[1]NitroMed Inc, 125 Spring Street, Lexington, MA 02421, USA; [2]BiogenIdec, 14 Cambridge Center, Cambridge, MA 02142, USA; [3]Surface Logix Inc, 50 Soldiers Field Place, Brighton, MA 02135, USA

Introduction

Since the publication of the first edition of "*In Vivo* Models of Inflammation" in 1999, evidence has continued to accumulate suggesting that asthma is predominantly a chronic inflammatory condition of the lower airways, characterized by varying degrees of airway obstruction or hyperresponsiveness and long-term ultrastructural abnormalities that may mitigate the effectiveness of some of the currently available therapies [1, 2]. These structural changes, which include airway edema, airway epithelial sloughing, increased airway wall thickening (smooth muscle hypertrophy and mucus gland hyperplasia), extracellular matrix abnormalities, and increased airway vascularity all contribute to airway remodeling and fibrosis. The dominant inflammatory cells responsible for initiating or propagating the aforementioned airway pathophysiology and subsequent ultrastructural changes include the T cell (Th2), mast cell, eosinophil, basophil, and macrophage [2, 3].

As alluded to in the previously published chapter on *in vivo* asthma models, much of this evidence continues to be assimilated using various *in vivo* animal models (from mouse to primate), and the caveat that these models are not replicas of the human disease and may not be predictive of human outcome still holds. Nonetheless, there has been a relative explosion in areas such as genomics, bioinformatics, and molecular pharmacology, which has helped us to better understand these various animal models and apply this understanding to the human condition.

This chapter highlights and pays particular attention to the following with respect to animal asthma models: (1) introduction of any new models or procedural changes within existing models, (2) new insights into pathogenesis or pathophysiology provided by the respective animal models of asthma, and (3) novel pharmacological approaches or new drugs tested in these models.

It should be noted that in many of the rodent studies summarized below, airway reactivity or bronchial responsiveness is commonly measured utilizing enhanced pause (Penh). The use of Penh as a measure of airway resistance (in the absence of

In Vivo Models of Inflammation, Vol. II, edited by Christopher S. Stevenson, Lisa A. Marshall and Douglas W. Morgan
© 2006 Birkhäuser Verlag Basel/Switzerland

direct measurement of this variable) remains a controversial subject [4], and is considered beyond the scope of this chapter.

Mouse models

The growing use of the mouse asthma model is likely due to the expanding availability of specific molecular, immunological and genetic tools to more completely explore and delineate possible inflammatory mechanisms contributing to the underlying pathophysiology. Excellent reviews [5–7] systematically outline the strengths and weaknesses associated with the mouse model, some of which are summarized in Table 1.

There now appears to be a mounting trend to use naturally occurring, airborne sensitizing allergens [e.g., house dust mite (HDM), ragweed, cockroach antigen, etc.] as opposed to ovalbumin (OVA) in the mouse model because these allergens are also known to precipitate asthma in humans [5, 6]. Repeated (10 day) intranasal exposure of BALB/c mice to purified HDM (*Dermatophagoides pteronyssinus*) in the absence of adjuvants produced an non-tolerant inflammatory response characterized by: (1) airway accumulation of eosinophils, Th2 lymphocytes, macrophages, and dendritic cells; (2) serum IgE increases; (3) mucous hypersecretion; and (4) airway hyperreactivity (AHR) that appeared to be partially mediated by GM-CSF [8]. Kim et al. [9] recently examined the effect of dexamethasone in a BALB/c mouse model that utilized a HDM extract composed of high concentrations of cockroach allergens for both intraperitoneal sensitization (day 0) and intratracheal challenge (days 14 and 21). Dexamethasone both prevented and reversed AHR and inflammation in this study. Interestingly, Fattouh et al. [10] have demonstrated that co-sensitization with intranasal HDM and aerosol OVA daily for 5 weeks produces a dramatic non-tolerant allergic response to aerosolized OVA characterized by eosinophilia, AHR, and elevation in splenocyte-derived Th2-cytokines (IL-4, IL-5, and IL-13). They suggest that HDM allergen may act in a number of ways to condition or alter the immunological environment, including airway epithelial disruption, promotion of a Th2-polarized cytokine responses, and direct proteolytic actions. These studies suggest that mouse asthma models developed using (multiple) environmental inhaled allergens may produce a non-tolerant pathophysiological response comparable to (if not better than) OVA that may be pharmacologically modified by reference or therapeutically novel classes of compounds. In an extension of these types of studies, investigators have now demonstrated that exposure to diverse external stimuli such as cigarette smoke [11–13], respiratory infection via *Mycoplasma pneumoniae* [14], and diesel exhaust [15] can exacerbate allergen-induced airway hyperresponsiveness, airway inflammation and remodeling.

In addition to what influence the types of sensitizing allergens has on the mouse model, there appears to be a large pathogenetic distinction between use of "acute"

Table 1 - Some comparison of human asthma and mouse asthma models

Characteristic	Human	Mouse
Airway inflammation	Sustained eosinophilia noted	Degree of eosinophilia variable; lack of degranulation and few eosinophils in epithelium; may be influenced by strain utilized and length of allergen exposure
Airway remodeling	Evidence of epithelial/goblet cell hyperplasia, subepithelial fibrosis, airway wall thickening	Mouse strain-, sensitization-, and challenge-dependent remodeling as per human noted; requires prolonged intermittent or continuous allergen exposure
Bronchoconstriction, airway hyperreactivity (AHR) and late phase response (LPR)	Sustained AHR and LPR (noted during asymptomatic periods)	Transient AHR and weak LPR; AHR requires high doses of agonist; unresponsive to many of the mediators associated with pathogenesis of asthma
Cytokines involvement	IL-4, IL-5, IL-13	Similar to human condition
Disease onset and progression	Spontaneous	Induced
Immunology	IgE/Th2-mediated	Mix of IgE/IgG and Th1/Th2-mediated events depending on mouse strain, sensitization, and challenge utilized; requires systemic immunization to see IgE mediation
Mast cell	Present in airways	Not noted in respiratory mucosa

versus "chronic" challenge protocols based upon the route of sensitization and challenge, concentration and duration of allergen challenge as well as differences between mouse strains utilized. The former are generally used to study aspects of Th2-mediated immune responses, hyperreactivity, airway inflammation, and mucus hypersecretion associated with the model, while the later (although there is evidence of immunological tolerance) may be useful in examining long-term airway remodeling, which is a growing clinical concern. A study reported by Shinagawa and Kojima [16] suggests that chronic allergen instillation (OVA; 3 days/week for up to 12 weeks) but not allergen inhalation (OVA; 5 days/week for up to 4 weeks) produced hallmarks of remodeling (marked airway wall thickening, mucous cell hypertrophy, airway eosinophilia, and collagen deposition) in the A/J mouse strain but to a lesser degree or not at all in BALB/c, C57BL/6, and C3H/HeJ mice. Note that in an acute inhaled-allergen challenge model (OVA; 1.0% for 1 h), the A/J mouse strain exhibited no allergen-induced increases in airway responsiveness to methacholine in the presence of mediocre bronchoalveolar lavage (BAL) accumulation of leukocytes (eosinophils and neutrophils) and cytokine generation [17], suggesting that route and total number of exposures may in fact regulate the pathophysiological response of the airways. Even in less intense chronic mouse models (OVA inhalation or instillation for 5–7 days following sensitization), evidence of both peribronchial and perivascular remodeling, including increased smooth muscle mass, collagen expression, and proliferation of epithelial and endothelial cells [18] as well as subepithelial matrix deposition of collagen and proteoglycans [19], have been noted. Overall, the "chronic" mouse models may be useful systems to examine the pharmacology of airway remodeling, and to gain insight into the corresponding human condition.

The allergic mouse continues to be routinely used as a model to explore the efficacy of various antibodies (primarily against cytokines and chemokines) as a therapeutic approach to treat the inflammatory asthmatic process. Some of these studies are summarized in Table 2 [20–24].

Genetically altered mice [severe combined immunodeficiency (SCID), transgenics including knockouts and knockins, etc.] are being increasingly used to delineate the relevance of various mechanisms in the pathobiology of asthma. The use of the SCID mouse, lacking both functional T and B lymphocytes, reconstituted with "sensitized" human peripheral blood mononuclear cells (hPBMC) is allowing investigators to explore components of the human immunobiology of asthma and use reagents especially chemokines and cytokines with limited species cross-reactivity. Duez et al. [25] demonstrated that SCID mice reconstituted intraperitoneally with hPBMC from HDM-sensitive patients and aerosol challenged daily for 4 days with HDM extract exhibited evidence of AHR, increases in human IgE in murine serum, increases in BAL fluid (BALF) IL-5 (but not IL-4 or tumor necrosis factor-α), and human pulmonary infiltrates but no lung eosinophilia. In an extension of this study, Tournoy et al. [26] showed that SCID mice reconstituted intratracheally with hPBMC from non-allergic donors could be driven from a Th1 to a Th2 phenotype (increases in lym-

Table 2 - Representative studies examining the effect of various cytokine or chemokine antibodies in the mouse asthma model

Study	Antibody/route of administration	AHR	Cell infiltration BALF/airway	Inhibition			Other
				BALF cytokines	Mucus	Remodeling	
Justice et al. [20]	CCR3/systemic + aerosol	+	+ EOS	−	+	ND	
Chung et al. [21]	CCR8 (TCA-3)/intranasal	ND	−	−	ND	ND	
Kumar et al. [22]	IL-5, IL-13 or IFN-γ intraperitoneal	+ (IL-13, IFN-γ)	+ EOS (IL-5, IL-13, IFN-γ)	ND	+ (IL-13)	+ collagen (IL-5)	
Cheng et al. [23]	IL-9/intravenous	+	+ EOS, + LYMPH + NEUT	+ IL-4, + IL-5, + IL-13	ND	ND	
Yang et al. [24]	IL-13/intravenous	+	+EOS +LYMPH	+IL-4 +IL-5 +TNF +Eotaxin	+	+Fibrosis	+MMP-9

+, positive effect; −, no effect; BALF, bronchoalveolar lavage fluid; EOS, eosinophils; LYMPH, lymphocytes; NEUT, neutrophil; MMP-9, matrix metalloproteinase-9; ND, not determined.

phocyte-derived IL-4 and IL-5 and decreases in IFN-γ upon intraperitoneal HDM plus adjuvant injection and a subsequent 19-day aerosol exposure to HDM. These animals exhibited AHR and pulmonary infiltration by human T lymphocytes in the absence of eosinophilia, but unlike the previous study exhibited no increases in human IgE in the murine serum. Adjuvant appeared to be necessary to drive the lymphocytes from a Th1 to a Th2 phenotype and elicit AHR. Interestingly, AHR in this human-mouse chimera model was blocked using anti-IL-4/IL-13 (DM-IL-4) or anti-IL-5 (TRFK-5) cytokine therapy, although allergen-induced Th2 cytokine production was not altered. Finally, SCID mice reconstituted intraperitoneally with hPBMC from HDM-sensitive donors primed with human HDM-pulsed dendritic cells and then challenged with HDM for 5 days developed intense pulmonary inflammation consisting of human lymphocytes and mouse eosinophils, increases in BALF IL-4 and IL-5, and elevations in human IgE in murine serum [27]. Based on the above examples, the humanized-SCID mouse asthma model will allow researchers to continue to explore the efficacy of novel therapeutics and mechanisms associated with asthma that may translate directly to the human condition.

The most significant growth in the past 6 years has been noted in terms of the use of transgenic mouse models of asthma especially in the area of cytokine research. An in-depth description of this area is beyond the scope of this chapter; however, some recent studies of interest are summarized below. To date, the use of IL-5 receptor null mice ($^{-/-}$) or monoclonal antibodies have produced equivocal results in establishing the precise role of IL-5 in the pathogenesis of asthma-associated eosinophilia and AHR [28]. Recent investigations, using IL-5 transgenic mice (to augment the eosinophilic response), have demonstrated that subchronic allergen-induced marked airway eosinophilia prevented AHR [29], while chronic allergen-induced eosinophilia enhanced subepithelial and peribronchial fibrosis [30], both through a transforming growth factor-β1 (TGF-β1) effect. The differences in these studies may reflect in part the aforementioned dichotomy associated with using different challenge protocols. In a further modification of the transgenic approach, Shen et al. [31] utilized intratracheal adoptive transfer of allergic eosinophils to eosinophil deficient IL-5$^{-/-}$ mice to demonstrate a probable interdependent relationship between CD4$^+$ T cells and eosinophils to initiate allergen-induced elevations in BALF cytokines, AHR, and mucus accumulation. Using both constitutive and inducible overexpression, as well as on/off triple transgenic mice, Elias et al. [32] have shown that IL-13-induced lung inflammation, characterized by airway remodeling, mucus hypersecretion, and fibrosis, is highly regulated via adenosine, chemokine receptor-2, matrix metalloproteinases, vascular endothelial growth factor, TGF-β1, and IL-11. Using IL-10$^{-/-}$ [33] and nitric oxide (NO) synthase2$^{-/-}$ mice [34], it has been recently suggested that NO may have a homeostatic role against allergen-associated immunopathobiology and AHR. There is still a great deal of speculation as to which NO synthase (1, 2 or 3) is bronchoprotective or anti-inflammatory. There is some evidence in transgenic mice overexpressing NOS-2 (the

CC10-rtTA-NOS-2 mouse) suggesting that this NOS is not proinflammatory, and may in fact reduce airway reactivity or provide bronchodilatory NO [35]. Finally, through the use of matrix metalloproteinase (MMP) -9 or -12 knockout mice, various investigators [36–38] have established that these MMPs play a critical role in the pathogenesis of allergen-induced airway inflammation possibly by modulating dendritic cell and T cell trafficking. These aforementioned transgenic-based mouse asthma model systems plus others exploring diverse therapeutic targets such as complement [39], tyrosine kinase inducible T cell kinase [40], NF-κB [41] and IL-17F [42] may provide excellent insight into the complex pathways that regulate asthma as an inflammatory disease.

Although a preponderance of work using the mouse asthma model has centered on transgenic applications and use of antibodies as therapeutic approaches or proof of concept testing, various novel classes of compounds of therapeutic interest have been evaluated since the last edition. The oral efficacy of the selective phosphodiesterase inhibitor, roflumilast (5 mg/kg/day), has been assessed in a chronic mouse model of asthma [43]. Roflumilast, similar to dexamethasone, appears to reduce airway inflammation, subepithelial fibrosis and epithelial hypertrophy. Like the MMPs mentioned above, there seems to be mounting evidence to support the importance of the mast cell-derived protease, tryptase, in the evolution of asthmatic inflammation. Oh et al. [44] have explored the role of tryptase in the mouse asthma model using the orally active reversible tryptase inhibitor, MOL 6131, and shown that it effectively reduced BALF and lung tissue inflammation as well as goblet cell hyperplasia and hypersecretion. Interestingly, when the compound was administered intranasally, the inhibitory effects were more pronounced.

CpG oligonucleotides acting as immunomodulatory agents via the Toll-like receptor (TLR)-9 to ultimately stimulate Th1 cytokine protective effects have also been critically examined. Kline et al. [45] and Jain et al. [46] demonstrated that CpG oligonucleotides may reverse or prevent acute and chronic allergen-induced inflammation and AHR through redirection from a Th2 to a Th1 microenvironment. This may be ultimately mediated by the up-regulation of IL-12 [47]. The importance of the Th1/Th2 balance and IL-12 up-regulation in asthma is further supported by the fact that intraperitoneal treatment with the TLR-2 agonist, PAM3CSK4, reduced lung inflammation, airway hyperresponsiveness, and serum IgE while increasing T-cell or dendritic cell derived IFN-γ, IL-12, and IL-10 in the mouse asthma model.

The mitogen-activated protein kinase (MAPK) family also appears to play an important immunomodulatory role in the mouse lung. Using repeat dosing (7 days) of aerosolized p38α MAPK antisense oligonucleotide, Duan et al. [48] demonstrated that this therapeutic approach effectively reduced allergen-induced eosinophilia, mucus hypersecretion, and AHR. The potential importance of this mechanism is also supported by the findings of Underwood et al. [49], who have demonstrated that the specific p38 MAPK inhibitor, SB 239063, administered orally, essentially abolished allergen-induced BALF eosinophilia in the allergic mouse model. Likewise, the combined MAPK/extracellular signal-regulated kinase (ERK) inhibitor, U0126, inhibited allergen-induced lung eosinophilia, increases in BALF IL-4, IL-5, IL-13, and eotaxin, mucus secretion, and airway hyperresponsiveness in the mouse when given via the intraperitoneal route [50].

Guinea pig models

It was noted in the last edition that the allergic guinea pig model has not been substantially altered since its inception some 90 years ago. This is again the case in the intervening years, although there have been modifications as investigators attempt to reproduce the different syndromes that make up human asthma using the guinea pig model. This includes a toluene-2,4-diisocyanate (TDI)-induced model which represents occupational asthma [51]. In this model, guinea pigs were first percutaneously sensitized with TDI followed by five tracheal challenges with a TDI mist. Both early and late responses were seen with only the late phase being inhibited by a corticosteroid. Nishitsuji et al. [52] developed a model of cough variant asthma. Animals were sensitized to OVA and the bronchial responsiveness as well as the cough reflex response to capsaicin were measured 72 h after an OVA inhalation challenge. Studies in humans indicate that exercise-induced asthma (EIA) worsens with increased dietary sodium. Studies in guineas pigs show that hyperpnea-induced airway obstruction, a model for EIA, is worsened in animals fed a high salt diet [53], and that this heightened response was inhibited by leukotriene blockade. This suggests that this guinea pig model has utility to mimic the human condition. Investigators have also examined the effects of different challenges in animals sensitized to OVA. Smith and Johnson [54] demonstrated that 5'-adenosine monophosphate produced a late asthmatic response (LAR) and an AHR in sensitized animals thus mimicking what is seen in atopic individuals. Studies in sensitized guinea pigs challenged with ultrasonically nebulized distilled water indicate a role for neurokinin (NK)-1 receptors in the resultant bronchoconstriction [55].

Probably the most striking change seen since the last edition is the number and class of compounds that have been evaluated in guinea pig models of asthma. There has been a continued interest in the efficacy of new phosphodiesterase (PDE) inhibitors, particularly those specific for PDE4 [56–58]. SCH 351591 and its active metabolite SCH 365351 both inhibited AHR and allergen-induced eosinophilia in the guinea pig [57]. Santing et al. [58] report that allergen-induced bronchoconstriction and AHR can be inhibited by PDE4 inhibitors, whereas inhibition of allergen-induced eosinophilia requires inhibition of both PDE3 and PDE4. There has also been increased interest in immunosuppressives, particularly by inhalation, for the treatment of asthma. Both cyclosporine [59] and FK-506 [60] given by inhalation inhibit AHR and cellular influx following allergen challenge. MX-68, a derivative of methotrexate, given orally inhibits the early and late bronchoconstrictive response as well as cellular influx [61]. Growing evidence that different prostanoids play either a pro- or anti-inflammatory role in the airways have prompted the synthesis of various new agents. The stable PGE_2 mimetic, misoprostol inhibits both bronchospasm and eosinophilia to an inhaled antigen challenge [62]. The PGD_2 receptor antagonist, S-5751 inhibits eosinophilia [63], whereas the combined TXA_2 synthase inhibitor, 5-LO inhibitor and antihistamine F-1322 inhibits the LAR and eosinophilia in an *Ascaris* model [64]. Investigators have also sought anti-inflammatory effects of recently approved drugs that are thought of as having primarily a bronchodilatory action. Thus, the leukotriene antagonist montelukast has been shown both to inhibit eosinophil influx [65] and to produce apoptosis of eosinophils [66]. The long-acting beta agonist, formoterol, and it R-R isomer have been shown to prevent eosinophilia and to protect against bronchospasm [67]. Interestingly, the long-acting muscarinic antagonist, tiotropium, has been shown to inhibit airway remodeling in a 12-week repeated challenge model [68]. Miscellaneous compounds that have shown efficacy in guinea pig models include ebselen [69], VUF-K-8788 [70], the tyrosine kinase inhibitor, Genistein [71] and the bradykinin B2 receptor antagonist FK-3657 [72].

The guinea pig has continued to be used to investigate the pathophysiology of asthma. The role of neurokinins and their receptors has been extensively studied with evidence for a complex interplay between the three subtypes (NK-1, NK-2, NK-3). NK-1 receptors contribute to AHR and eosinophil, neutrophil and lymphocyte infiltration [73], whereas NK-2 receptors have been reported to mediate late phase hyperreactivity, late phase bronchoconstriction and the influx of neutrophils and lymphocytes but not eosinophils [73, 74]. NK-3 receptors have also been reported to mediate AHR in a severe asthma model [75, 76]. Arginase upregulation has also been reported to contribute to AHR, presumably by lowering NO synthesis from constitutive NO synthases by competing for their common substrate, L-arginine [77]. The complement fragment C3a is also implicated in the bronchoconstrictor response to allergen as guinea pigs with a natural deficiency in the C3a receptor have a reduced response compared to wild-type animals [78]. The ability of an IL-13 binding fusion protein to inhibit eosinophilia and AHR provides further support for a role of IL-13 in asthma pathophysiology [79].

Rat models

The Brown Norway (BN) strain continues to be the primary strain used and the sensitization procedures follow those outlined in the last edition. As in the guinea pig, investigators have also used chemicals, known to produce occupational asthma, as sensitizing and challenge agents in the rat. Thus, both diphenylmethane-4,4'-diisocyanate (MDI) [80] administered dermally or by inhalation [80], and trimellitic anhydride (TMA) administered dermally [81], have been shown to produce IgE titers and to produce bronchoconstriction and eosinophilia. Dong et al. [82] used HDM antigen with or without the systemic administration of *Bordetella pertussis* in 3-week-old BN rats to establish a model that would more closely reflect the developing immune system of children. They showed that the sensitization process was enhanced when both agents were administered simultaneously. The Flinders sensitive line of rats, which are hyperresponsive to cholinergic stimuli, show enhanced bronchoconstrictor responses and airway inflammation when they are sensitized to OVA and subsequently challenged [83], indicating that neural pathways may play an important role in the asthma phenotype.

The rat model has also been increasingly used to examine the phenomenon of airway remodeling in the pathogenesis of asthma, as well as potential therapeutic interventions to prevent remodeling. As with many models, the parameters used to elicit remodeling vary considerably between investigators, but one constant is the need to repeatedly challenge the animals with inhaled antigen. Chung's group at the University of London used six OVA challenges every 3rd day [84–86] to elicit remodeling and measured goblet cell hyperplasia, epithelial cell proliferation and airway smooth muscle (ASM) proliferation as indices of the remodeling process. ASM and epithelial proliferation were studied by measuring 5-bromo-2'-deoxyuridine. Using this chronic antigen-challenge paradigm, corticosteroids [84] and a Jun N-terminal kinase (JNK) inhibitor (SP600125) [85] have been shown to inhibit airway remodeling. Xu et al. [87] examined ASM hyperplasia in animals sensitized with OVA with *B. pertussis*, and subsequently challenged with OVA given by inhalation on three occasions, 5 days apart. This study showed a modest increase in ASM cells in the antigen-challenge group compared to the saline-challenge group. Vanacker et al. [88] report on a model where animals were challenged for 28 days with antigen and noticed an increase in goblet cell numbers, epithelial cell proliferation, airway wall area, fibronectin deposition and collagen deposition. These structural changes were inhibited by fluticasone administered daily for the last 2 weeks of the challenge period.

Due to its relevance and utility as a toxicology species, and thus the ability to determine the therapeutic window in the same species, rat models have continued to be used to study the potential of new therapeutic compounds. As noted in the guinea pig, the rat has been used to investigate novel PDE4 inhibitors including NVP-ABE171 [89] and YM976 [90]. The rat has also been used to examine whether parenteral administration of compounds to the lung would provide efficacy with a better safety profile than oral administration. Compounds studied include the "soft" steroids cilcesonide [51] and BNP-166 [91], as well as the macrolide MLD987 [92]. In addition, gene-based therapies have also been studied in the rat, including the use of gene therapy with Galectin 3 [93] and antisense to Syk kinase [94]. As asthma has been increasingly recognized as a chronic disease, mediated via a Th2 imbalance, a considerable number of immunomodulators have been examined for efficacy in the rat asthma model. These can be found in Table 3 [95–105].

Table 3 - The effect of various immunomodulators in the rat asthma model

Compound	Class of compound	Acute/chronic admin.	BHR	Leukocyte infiltration	Other	Ref.
Compound A	IKK β kinase inhibitor	C	Yes	Yes		[95]
SP600125	JNK kinase inhibitor	A	No	Yes		[96]
YM-90709	IL-5 antagonist	?	–	Yes		[97]
SAR943	Immunomodulator	A	No	Yes		[98]
SAR943	Immunomodulator	C	–	No	ASM	[99]
IMM125	Immunomodulator	A	No	Yes	Cytokines	[98]
Suplatast	IgE inhibitor	C	–	Yes	Cytokines	[100]
CGS 21680	Adenosine A2a agonist	A	–	Yes		[101]
SP100030	Transcription factor inhibitor	SubC	–	-	Lymphocytes	[102]
TEI-9874	IgE inhibitor	A	–	Yes	EAR and LAR	[103]
PS-519	Proteasome inhibition	A?	–	Yes		[104]
Leflunomide	Immunomodulator	C	–	-	IgE levels	[105]

–, not examined.

11

Rabbit models

In comparison with other species, rabbit models have undergone less evolution since the last edition. In an interesting study the hypothesis that systemic allergy and asthma worsens the outcome of cardiovascular complications was examined [106]. The authors found that rabbits sensitized and subsequently challenged by aerosol have increased infarct size and neutrophil infiltration in a model of myocardial ischemia-reperfusion injury compared to non-sensitized rabbits [106]. In a model of gastroesophageal acid reflux, Gallelli et al. [107] demonstrated that the bronchoconstriction elicited by intraesophageal instillation of HCl was mediated via tachykinins acting on NK-1 and NK-2 receptors. Hogman et al. [108] reported that both inhaled histamine and hypertonic saline increase airway reactivity in non-sensitized rabbits and noted that asthmatics have increased airway reactivity after nebulizing hypertonic saline.

The rabbit has continued to be used to investigate the pathophysiology of asthma particularly by investigators using isolated muscle preparations. Grunstein's group has investigated the effects of HDM allergen [109], rhinovirus [110] and IL-1β [111] on contractile and relaxant responses in ASM isolated from the rabbit. They showed that the HDM allergen Der p 1 enhances contractile responses and inhibits relaxant responses via activation of ERK1/2 pathways, and that this is regulated by p38 MAP kinase signaling [109]. Rhinovirus affects ASM responsiveness by an ICAM-1-dependent activation of IL-5 pathways, which in turn releases IL-1β from the ASM [110]. Using *in vivo* models, investigators have examined the role of integrins [112], adenosine [113], tachykinins [114], bradykinin [114] and reactive oxygen species [115] on antigen-induced pulmonary responses. Gascoigne et al. [112] report that VLA4 is involved in acute bronchoconstriction in the rabbit, whereas eosinophil recruitment and infiltration involves VLA4 and LFA-1, and lymphocyte recruitment involves LFA-1 and Mac-1 [112]. Further evidence for a role of adenosine in the allergic response in the rabbit comes from studies using a new selective adenosine A1 receptor antagonist L-97-1 [113]. Other studies have shown a mixed role for bradykinin, neurokinins and reactive oxygen species. Thus, superoxide dismutase inhibits AHR, but has no effect on airway inflammation in a chronic model [115], whereas neurokinins acting via NK-2 receptors inhibit acute bronchoconstriction to antigen, but have no effect on the resultant eosinophilia or AHR [114]. Bradykinin B2 receptors appear to play no role in the airway response to allergen challenge [114].

Canine models

In terms of canine models of asthma, the focus since the first edition of "*In Vivo Models of Inflammation*" published in 1999 has centered on: (1) continued development and use of neonatally ragweed-sensitized beagles from allergic parents and (2) use of a dry air challenge model to examine exercise-induced asthma. Although not reviewed in detail here, a ragweed-sensitized dog model of allergic rhinitis has been developed and appears to be useful to study the antiallergic properties of new therapeutic agents [116, 117].

Beagle puppies of allergic, high serum IgE parents sensitized with intraperitoneal ragweed at 24 h post-partum to 22 weeks of age and subsequently exposed to multiple aerosol ragweed challenges exhibited evidence of elevated serum IgE, AHR to histamine, methacholine, and neurokinin A as well as BALF eosinophilia [118–120]. T lymphocytes removed from these animals 4 h following segmental ragweed challenge exhibited evidence of localized activation. In general, while this canine asthma model appears to exhibit many of the characteristics of the human asthmatic condition, it has not been routinely utilized perhaps because of the difficulties associated with housing and maintenance of this particular model.

Hyperventilation dry air challenge of anesthetized dogs can elicit bronchoconstriction and BALF eicosanoid (leukotriene C_4 and E_4; prostaglandin D_2, $F_2\alpha$, and thromboxane B_2) generation [121, 122]. These pathophysiological changes could be attenuated by aerosolized heparin. Using a repeat cold dry air challenge model of hyperpnea, Davis et al. [123] demonstrated evidence of transient airway remodeling characterized by epithelial cell hypertrophy, thickened lamina propria, and tissue accumulation of eosinophils, neutrophils, and mast cells.

Sheep models

The sheep lung exhibits numerous physiological and pathophysiological similarities to humans with respect to lung size, anatomy and development, bronchial circulation and airway innervation, characteristics of mast cells and mucus production, and high serum IgE and allergic inflammation in the lungs after allergen challenge. In addition, they are responsive to bronchospastic agents and modulators that are effective in humans. Two sheep models are currently available, the first and most widely utilized to date using animals having a natural skin sensitivity to *Ascaris suum* [124, 125]. On exposure to aerosolized antigen, an early bronchoconstriction and pulmonary hyperinflation are observed in these animals. A proportion of these early responders called dual responders go on to develop bronchoconstriction 7–8 h later and a nonspecific AHR at 24 h, which can last up to 2 weeks. In addition, lung inflammation characterized 24 h later by BAL and/or tissue accumulation of macrophages, neutrophils, eosinophils, and lymphocytes is evident and correspondingly more pronounced in dual responders. As mentioned in the previous edition, the introduction of a technique (lavage via a double-balloon nasotracheal tube) to isolate and study upper airway epithelial function (i.e., mucus secretion) and inflammation in this model has greatly expanded our understanding of this model [126].

Corticosteriods [127], mast cell modifying agents [128, 129], and heparin [130] are particularly effective at reducing the allergic response in sheep, implicating a significant role for mast cells and their mediators in this model. Neutrophils appear to be important in both early and dual responders, whereas eosinophils are found in increased numbers in the BAL and airway wall of dual responders only [124, 125]. The inflammatory component, early/late phase response and AHR can be inhibited by a wide variety of modulators, which suggests contributing roles for adhesion molecules [124, 125], eicosanoids [131, 132], and tissue kallikrein/kinins [133].

An alternative sheep model involves the use of the more appropriate sensitizing antigen, HDM [134]. This model utilizes a systemic sensitization followed by a local lung challenge approach followed by evaluation of subsequent pulmonary responses. This allows for a more controlled evaluation of functional, cellular and immune responses. At 48 h following challenge there are significant changes in blood, BAL and tissue eosinophils and activated CD4+ cells, not dissimilar to the situation that occurs in humans [134]. The model responds to some standard clinically used asthma drugs but novel therapies have yet to be tested. The model has also been modified using repeated intratracheal challenge with HDM for 6 months to look at the relationship between airway remodeling and inflammation [135]

Despite the limited availability of this preparation (i.e., use by a small group of laboratories), the allergic sheep model continues to be extensively utilized to demonstrate intravenous, oral or inhaled efficacy of a wide range of compounds. Some of the more important studies generated since the publication of the first edition of "*In Vivo* Models of Inflammation" in 1999, are summarized in Table 4 [125, 136–141].

Primate models

The rationale for developing primate models of asthma reflects the genetic, physiological, and immunological similarity to humans as well as the similarity of human and monkey lungs in terms of anatomy, histology, and ultrastructure. Their size also allows for pulmonary function testing using non-invasive forced oscillation techniques [142] and bronchoscopy similar to that used in humans [143]. Most of the models developed have used rhesus, cynomolgus or squirrel monkeys, which are naturally sensitive to *A. suum*, presumably due to the prior sensitization by nematode infestation [144]. In 2001, aerosolized dust mite was used to induce progressive inflammation, decline in pulmonary function and structural changes in the airway walls of rhesus monkeys [143]. Although this model closely resembles the human disease, it is labor intensive and the practicality for drug screening is limited. Recently, a more practical model of HDM-induced asthma has been developed, whereby animals are sensitized from birth with allergen [145] that can be used for drug screening and research [146].

Table 4 - Representative studies examining the effect of various compounds on antigen-induced responses in the sheep asthma model

Compound	Route of administration	Inhibition						Ref.
		EPR	LPR	AHR	PMN	EOS		
α4 integrin antagonist BIO-1211	aero./i.v.	Yes	Yes	Yes	Yes	Yes		125
α4 integrin antagonist non-peptides	aero.	Yes	Yes	Yes	–	–		136
α1 integrin antibody mAb AQC2	aero.	No	Yes	Yes	No	Yes		137
Tryptase inhibitor RWJ-56423	aero.	Yes	Yes	Yes	–	–		128
L-Selectin antibody MECA-79	i.v.	No	Yes	Yes	Yes	Yes		138
Sterol derivative IPL576,092	aero.	Yes	Yes	Yes	–	–		139
IL-5 synthesis inhibitor	i.t.	–	Yes	Yes	–	–		140
Chemotaxis inhibitor TAK-661	aero.	No	Yes	–	–	Yes		141

i.v., intravenous; p.o., oral; aero., aerosol; i.t. intratracheal; EPR, early phase response; LPR, late phase response; AHR, airway hyper-reactivity; PMN, pulmonary neutrophilia; EOS, pulmonary eosinophilia; –, not tested or not shown.

Table 5 - Representative studies examining the effect of various compounds on antigen-induced responses in the monkey asthma model

Compound	Route of administration	Inhibition					Ref.
		EPR	LPR	AHR	PMN	EOS	
Anti-human IL-5 mAb SB-240563	i.v.	No	–	–	–	Yes	147
Immunostimulatory oligonucleotide	aero.	–	–	Yes	–	Yes	148
K-ATP channel modulator KCO912	aero.	–	–	Yes	–	–	149
Phosphodiesterase 4 inhibitor SCH 351591	p.o.	No	–	–	–	Yes	150

i.v., intravenous; p.o., oral; aero., aerosol; i.t., intratracheal; EPR, early phase response; LPR, late phase response; AHR, airway hyper-reactivity; PMN, pulmonary neutrophilia; EOS, pulmonary eosinophilia; –, not tested or not shown.

Primate models of asthma have been used to evaluate both established and novel anti-allergy and anti-asthmatic compounds. Clearly cost can be prohibitive and, despite the tremendous potential of non-human primate models, they will not replace other species. However, monkey models have developed to a high degree of sophistication that allows drug discoverers and developers to confirm their findings in a species more relevant to man before entering into expensive and time consuming clinical trials. Some of the more recent potential therapeutics examined in the monkey are summarized in Table 5 [147–150].

Since the publication of the first edition of "*In Vivo* Models of Inflammation", it is evident that research examining the inflammatory basis of asthma continues to be fervently pursued. Surprisingly, relatively few new therapeutics have been introduced in this area in recent years. The use of some of the animal models of "asthma" discussed herein provide researchers with useful *in vivo* models to explore important components of the human asthmatic response and may foster the further development of novel, anti-inflammatory therapeutic approaches to treat this debilitating disease.

References

1   Homer RJ, Elias JA (2005) Airway remodeling in asthma: therapeutic implications of mechanisms. *Physiology* 20: 28–35

2   Fabbri L, Peters SP, Pavord I, Wenzel SE, Lazarus SC, Macnee W, Lemaire F, Abraham E (2005) Allergic rhinitis, asthma, airway biology, and chronic obstructive pulmonary disease in AJRCCM in 2004. *Am J Respir Crit Care Med* 171: 686–698

3   Amin K, Ludviksdottir D, Janson C, Nettelbladt O, Bjornsson E, Roomans GM, Boman G, Seveus L, Venge P (2000) Inflammation and structural changes in the airways of patients with atopic and nonatopic asthma. *Am J Respir Crit Care Med* 162: 2295–2301

4   Sly PD, Turner DJ, Hantos Z (2004) Measuring lung function in murine models of pulmonary disease. *Drug Discov Today* 1: 337–343

5   Torres R, Picado C, de Mora F (2005) Use of the mouse to unravel allergic asthma: a review of the pathogenesis of allergic asthma in mouse models and its similarity to the condition in humans. *Arch Bronconeumol* 41: 141–152

6   Epstein MM (2004) Do mouse models of allergic asthma mimic clinical disease? *Int Arch Allergy Immunol* 133: 84–100

7   Kumar RK, Foster PS (2002) Modeling allergic asthma in mice: pitfalls and opportunities. *Am J Respir Cell Mol Biol* 27: 267–272

8   Cates EC, Fattouh R, Wattie J, Inman MD, Goncharova S, Coyle AJ, Gutierrez-Ramos JC, Jordana M (2004) Intranasal exposure of mice to house dust mite elicits allergic airway inflammation via a GM-CSF-mediated mechanism. *J Immunol* 173: 6384–6392

9   Kim J, McKinley L, Siddiqui J, Bolgos GL, Remick DG (2004) Prevention and reversal of pulmonary inflammation and airway hyperresponsiveness by dexamethasone treat-

ment in a murine model of asthma induced by house dust. *Am J Physiol Lung Cell Mol Physiol* 287: L503–509

10 Fattouh R, Pouladi MA, Alvarez D, Johnson JR, Walker TD Goncharova S, Inman MD, Jordana M (2005) House dust mite facilitates ovalbumin-specific allergic sensitization and airway inflammation. *Am J Respir Crit Care Med* 172: 314–321

11 Moerloose KB, Pauwels RA, Joos GF (2005) Short-term cigarette smoke exposure enhances allergic airway inflammation in mice. *Am J Respir Crit Care Med* 172: 168–172

12 Barrett EG, Wilder JA, March TH, Espindola T, Bice DE (2002) Cigarette smoke-induced airway hyperresponsiveness is not dependent on elevated immunoglobulin and eosinophilic inflammation in a mouse model of allergic airway disease. *Am J Respir Crit Care Med* 165: 1410–1408

13 Seymour BW, Schelegle ES, Pinkerton KE, Friebertshauser KE, Peake JL, Kurup VP, Coffman RL, Gershwin LJ (2003) Second-hand smoke increases bronchial hyperreactivity and eosinophilia in a murine model of allergic aspergillosis. *Clin Dev Immunol* 10: 35–42

14 Chu HW, Rino JG, Wexler RB, Campbell K, Harbeck RJ, Martin RJ (2005) *Mycoplasma pneumoniae* infection increases airway collagen deposition in a murine model of allergic airway inflammation. *Am J Physiol Lung Cell Mol Physiol* 289: L125–133

15 Ichinose T, Takano H, Sadakane K, Yanagisawa R, Yoshikawa T, Sagai M, Shibamoto T (2004) Mouse strain differences in eosinophilic airway inflammation caused by intra-tracheal instillation of mite allergen and diesel exhaust particles. *J Appl Toxicol* 24: 69–76

16 Shinagawa K, Kojima M (2003) Mouse model of airway remodeling: strain differences. *Am J Respir Crit Care Med* 168: 959–967

17 Whitehead GS, Walker JK, Berman KG, Foster WM, Schwartz DA (2003) Allergen-induced airway disease is mouse strain dependent. *Am J Physiol Lung Cell Mol Physiol* 285: L32–42

18 Tormanen KR, Uller L, Persson CG, Erjefalt JS (2005) Allergen exposure of mouse airways evokes remodeling of both bronchi and large pulmonary vessels. *Am J Respir Crit Care Med* 171: 19–25

19 Reinhardt AK, Bottoms SE, Laurent GJ, McAnulty RJ (2005) Quantification of collagen and proteoglycan deposition in a murine model of airway remodelling. *Respir Res* 6: 30–43

20 Justice JP, Borchers MT, Crosby JR, Hines EM, Shen HH, Ochkur SI, McGarry MP, Lee NA, Lee JJ (2003) Ablation of eosinophils leads to a reduction of allergen-induced pulmonary pathology. *Am J Physiol Lung Cell Mol Physiol* 284: L169–178

21 Chung CD, Kuo F, Kumer J, Motani AS, Lawrence CE, Henderson WR Jr, Venkataraman C (2003) CCR8 is not essential for the development of inflammation in a mouse model of allergic airway disease. *J Immunol* 170: 581–587

22 Kumar RK, Herbert C, Webb DC, Li L, Foster PS (2004) Effects of anticytokine therapy in a mouse model of chronic asthma. *Am J Respir Crit Care Med* 170: 1043–1048

23 Cheng G, Arima M, Honda K, Hirata H, Eda F, Yoshida N, Fukushima F, Ishii Y, Fukuda T (2002) Anti-interleukin-9 antibody treatment inhibits airway inflammation and hyperreactivity in mouse asthma model. *Am J Respir Crit Care Med* 166: 409–416

24 Yang G, Li L, Volk A, Emmell E, Petley T, Giles-Komar J, Rafferty P, Lakshminarayanan M, Griswold DE, Bugelski PJ, Das AM (2005) Therapeutic dosing with anti-interleukin-13 monoclonal antibody inhibits asthma progression in mice. *J Pharmacol Exp Ther* 313: 8–15

25 Duez C, Kips J, Pestel J, Tournoy K, Tonnel AB, Pauwels R (2000) House dust mite-induced airway changes in hu-SCID mice. *Am J Respir Crit Care Med* 161: 200–206

26 Tournoy KG, Kips JC, Pauwels RA (2001) The allergen-induced airway hyperresponsiveness in a human-mouse chimera model of asthma is T cell and IL-4 and IL-5 dependent. *J Immunol* 166: 6982–6991

27 Hammad H, Lambrecht BN, Pochard P, Gosset P, Marquillies P, Tonnel AB, Pestel J (2002) Monocyte-derived dendritic cells induce a house dust mite-specific Th2 allergic inflammation in the lung of humanized SCID mice: involvement of CCR7. *J Immunol* 169: 1524–1534

28 Mathur M, Herrmann K, Li X, Qin Y, Weinstock J, Elliott D, Monahan J, Padrid P (1999) TRFK-5 reverses established airway eosinophilia but not established hyperresponsiveness in a murine model of chronic asthma. *Am J Respir Crit Care Med* 159: 580–587

29 Kobayashi T, Iijima K, Kita H (2003) Marked airway eosinophilia prevents development of airway hyper-responsiveness during an allergic response in IL-5 transgenic mice *J Immunol* 170: 5756–5763

30 Tanaka H, Komai M, Nagao K, Ishizaki M, Kajiwara D, Takatsu K, Delespesse G, Nagai H (2004) Role of interleukin-5 and eosinophils in allergen-induced airway remodeling in mice. *Am J Respir Cell Mol Biol* 31: 62–68

31 Shen HH, Ochkur SI, McGarry MP, Crosby JR, Hines EM, Borchers MT, Wang H, Biechelle TL, O'Neill KR, Ansay TL et al (2003) A causative relationship exists between eosinophils and the development of allergic pulmonary pathologies in the mouse. *J Immunol* 170: 3296–3305

32 Elias JA, Zheng T, Lee CG, Homer RJ, Chen Q, Ma B, Blackburn M, Zhu Z (2003) Transgenic modeling of interleukin-13 in the lung. *Chest* 123: 339S–345S

33 Ameredes BT, Sethi JM, Liu HL, Choi AM, Calhoun WJ (2005) Enhanced nitric oxide production associated with airway hyporesponsiveness in the absence of IL-10. *Am J Physiol Lung Cell Mol Physiol* 288: L868–873

34 Rodriguez D, Keller AC, Faquim-Mauro EL, de Macedo MS, Cunha FQ, Lefort J, Vargaftig BB, Russo M (2003) Bacterial lipopolysaccharide signaling through Toll-like receptor 4 suppresses asthma-like responses via nitric oxide synthase 2 activity. *J Immunol* 17: 1001–1008

35 Hjoberg J, Shore S, Kobzik L, Okinaga S, Hallock A, Vallone J, Subramaniam V, De Sanctis GT, Elias JA, Drazen JM, Silverman ES (2004) Expression of nitric oxide syn-

thase-2 in the lungs decreases airway resistance and responsiveness. *J Appl Physiol* 97: 249–259

36   Vermaelen KY, Cataldo D, Tournoy K, Maes T, Dhulst A, Louis R, Foidart JM, Noel A, Pauwels R (2003) Matrix metalloproteinase-9-mediated dendritic cell recruitment into the airways is a critical step in a mouse model of asthma. *J Immunol* 17: 1016–1022

37   McMillan SJ, Kearley J, Campbell JD, Zhu XW, Larbi KY, Shipley JM, Senior RM, Nourshargh S, Lloyd CM (2004) Matrix metalloproteinase-9 deficiency results in enhanced allergen-induced airway inflammation. *J Immunol* 172: 2586–2594

38   Warner RL, Lukacs NW, Shapiro SD, Bhagarvathula N, Nerusu KC, Varani J, Johnson KJ (2004) Role of metalloelastase in a model of allergic lung responses induced by cockroach allergen. *Am J Pathol* 165: 1921–1930

39   Drouin SM, Corry DB, Hollman TJ, Kildsgaard J, Wetsel RA (2002) Absence of the complement anaphylatoxin C3a receptor suppresses Th2 effector functions in a murine model of pulmonary allergy. *J Immunol* 169: 5926–5933

40   Mueller C, August A (2003) Attenuation of immunological symptoms of allergic asthma in mice lacking the tyrosine kinase ITK. *J Immunol* 170: 5056–5063

41   Poynter ME, Cloots R, van Woerkom T, Butnor KJ, Vacek P, Taatjes DJ, Irvin CG, Janssen-Heininger YM (2004) NF-kappa B activation in airways modulates allergic inflammation but not hyperresponsiveness. *J Immunol* 173: 7003–7009

42   Oda N, Canelos PB, Essayan DM, Plunkett BA, Myers AC, Huang SK (2005) Interleukin-17F induces pulmonary neutrophilia and amplifies antigen-induced allergic response. *Am J Respir Crit Care Med* 171: 12–18

43   Kumar RK, Herbert C, Thomas PS, Wollin L, Beume R, Yang M, Webb DC, Foster PS (2003) Inhibition of inflammation and remodeling by roflumilast and dexamethasone in murine chronic asthma. *J Pharmacol Exp Ther* 307: 349–355

44   Oh SW, Pae CI, Lee DK, Jones F, Chiang GK, Kim HO, Moon SH, Cao B, Ogbu C, Jeong KW et al (2002) Tryptase inhibition blocks airway inflammation in a mouse asthma model. *J Immunol* 168: 1992–2000

45   Kline JN, Kitagaki K, Businga TR, Jain VV (2002) Treatment of established asthma in a murine model using CpG oligodeoxynucleotides. *Am J Physiol Lung Cell Mol Physiol* 283: L170–179

46   Jain VV, Businga TR, Kitagaki K, George CL, O'Shaughnessy PT, Kline JN (2003) Mucosal immunotherapy with CpG oligodeoxynucleotides reverses a murine model of chronic asthma induced by repeated antigen exposure. *Am J Physiol Lung Cell Mol Physiol* 285: L1137–1146

47   Choudhury BK, Wild JS, Alam R, Klinman DM, Boldogh I, Dharajiya N, Mileski WJ, Sur S (2002) *In vivo* role of p38 mitogen-activated protein kinase in mediating the anti-inflammatory effects of CpG oligodeoxynucleotide in murine asthma. *J Immunol* 169: 5955–5961

48   Duan W, Chan JH, McKay K, Crosby JR, Choo HH, Leung BP, Karras JG, Wong WS (2005) Inhaled p38alpha mitogen-activated protein kinase antisense oligonucleotide attenuates asthma in mice. *Am J Respir Crit Care Med* 171: 571–578

49 Underwood DC, Osborn RR, Kotzer CJ, Adams JL, Lee JC, Webb EF, Carpenter DC, Bochnowicz S, Thomas HC, Hay DW, Griswold DE (2000) SB 239063, a potent p38 MAP kinase inhibitor, reduces inflammatory cytokine production, airways eosinophil infiltration, and persistence. *J Pharmacol Exp Ther* 293: 281–288

50 Duan W, Chan JH, Wong CH, Leung BP, Wong WS (2004) Anti-inflammatory effects of mitogen-activated protein kinase kinase inhibitor U0126 in an asthma mouse model. *J Immunol* 172: 7053–7059

51 Nabe T, Yamauchi K, Shinjo Y, Niwa T, Imoto K, Koda A, Kohno S (2005) Delayed-type asthmatic response induced by repeated intratracheal exposure to toluene-2,4-diisocyanate in guinea pigs. *Int Arch Allergy Immunol* 137: 115–124

52 Nishitsuji M, Fujimura M, Oribe Y, Nakao S (2004) A guinea pig model for cough variant asthma and role of tachykinins. *Exp Lung Res* 30: 723–737

53 Mickleborough TD, Gotshall RW, Rhodes J, Tucker A, Cordain L (2001) Elevating dietary salt exacerbates hyperpnea-induced airway obstruction in guinea pigs. *J Appl Physiol* 91: 1061–1066

54 Smith N, Johnson FJ (2005) Early- and late-phase bronchoconstriction, airway hyperreactivity and cell influx into the lungs, after 5'-adenosine monophosphate inhalation: comparison with ovalbumin. *Clin Exp Allergy* 35: 522–530

55 Ishiura Y, Fujimura M, Myou S, Amemiya T, Nobata K, Kurashima K, Nonomura A (2003) Airway eosinophil accumulation on sensory neuropeptide release in a guinea pig model of distilled-water-induced bronchoconstriction. *J Investig Allergol Clin Immunol* 13: 79–86

56 Kim E, Chun HO, Jung SH, Kim JH, Lee JM, Suh BC, Xiang MX, Rhee CK (2003) Improvement of therapeutic index of phosphodiesterase type IV inhibitors as anti-Asthmatics. *Bioorg Med Chem Lett* 13: 2355–2358

57 Billah MM, Cooper N, Minnicozzi M, Warneck J, Wang P, Hey JA, Kreutner W, Rizzo CA, Smith SR, Young S et al (2002) Pharmacology of N-(3,5-dichloro-1-oxido-4-pyridinyl)-8-methoxy-2-(trifluoromethyl)-5-quinoline carboxamide (SCH 351591), a novel, orally active phosphodiesterase 4 inhibitor. *J Pharmacol Exp Ther* 302: 127–137

58 Santing RE, de Boer J, Rohof A, van der Zee NM, Zaagsma J (2001) Bronchodilatory and anti-inflammatory properties of inhaled selective phosphodiesterase inhibitors in a guinea pig model of allergic asthma. *Eur J Pharmacol* 429: 335–344

59 Xie QM, Chen JQ, Shen WH, Yang QH, Bian RL (2002) Effects of cyclosporin A by aerosol on airway hyperresponsiveness and inflammation in guinea pigs. *Acta Pharmacol Sin* 23: 243–247

60 Morishita Y, Hirayama Y, Miyayasu K, Tabata K, Kawamura A, Ohkubo Y, Mutoh S (2005) FK506 aerosol locally inhibits antigen-induced airway inflammation in Guinea pigs. *Int Arch Allergy Immunol* 136: 372–378

61 Nagao K, Akabane H, Masuda T, Komai M, Tanaka H, Nagai H (2004) Effect of MX-68 on airway inflammation and hyperresponsiveness in mice and guinea-pigs. *J Pharm Pharmacol* 56: 187–196

62 Smith WG, Thompson JM, Souresrafil NS, McKearn JP (1998) Initial studies on the

effect of inhaled misoprostol in a guinea pig model of allergic bronchoconstriction. *Am J Ther* 2: 755–760

63   Arimura A, Yasui K, Kishino J, Asanuma F, Hasegawa H, Kakudo S, Ohtani M, Arita H (2001) Prevention of allergic inflammation by a novel prostaglandin receptor antagonist, S-5751. *J Pharmacol Exp Ther* 298: 411–419

64   Mochizuki A, Tamura N, Yatabe Y, Onodera S, Hiruma T, Inaba N, Kusunoki J, Tomioka H (2001) Suppressive effects of F-1322 on the antigen-induced late asthmatic response and pulmonary eosinophilia in guinea pigs. *Eur J Pharmacol* 430: 123–133

65   Leick-Maldonado EA, Kay FU, Leonhardt MC, Kasahara DI, Prado CM, Fernandes FT, Martins MA, Tiberio IF (2004) Comparison of glucocorticoid and cysteinyl leukotriene receptor antagonist treatments in an experimental model of chronic airway inflammation in guinea-pigs. *Clin Exp Allergy* 34: 145–152

66   Wu YQ, Zhou CH, Zhang HQ (2004) Effects of montelukast on apoptosis and Fas mRNA expression of eosinophils in airway of asthmatic guinea pigs *Yao Xue Xue Bao* 39: 769–773

67   Xie QM, Chen JQ, Shen WH, Yang QH, Bian RL (2003) Comparison of bronchodilating and antiinflammatory activities of oral formoterol and its (R,R)-enantiomers. *Acta Pharmacol Sin* 24: 277–282

68   Gosens R, Bos IS, Zaagsma J, Meurs H 2005) Protective effects of tiotropium bromide in the progression of airway smooth muscle remodeling. *Am J Respir Crit Care Med* 171: 1096–1102

69   Zhang M, Nomura A, Uchida Y, Iijima H, Sakamoto T, Iishii Y, Morishima Y, Mochizuki M, Masuyama K, Hirano K et al (2002) Ebselen suppresses late airway responses and airway inflammation in guinea pigs. *Free Radic Biol Med* 32: 454–464

70   Takizawa T, Watanabe C, Saiki I, Wada Y, Tohma T, Nagai H (2001) Effects of a new antiallergic drug, VUF-K-8788, on infiltration of lung parenchyma by eosinophils in guinea pigs and eosinophil-adhesion to human umbilical vein endothelial cells (HUVEC). *Biol Pharm Bull* 24: 1127–1132

71   Duan W, Kuo IC, Selvarajan S, Chua KY, Bay BH, Wong WS (2003) Antiinflammatory effects of genistein, a tyrosine kinase inhibitor, on a guinea pig model of asthma. *Am J Respir Crit Care Med* 167: 185–192

72   Hirayama Y, Miyayasu K, Yamagami K, Imai T, Ohkubo Y, Mutoh S (2003) Effect of FK3657, a non-peptide bradykinin B2 receptor antagonist, on allergic airway disease models. *Eur J Pharmacol* 467: 197–203

73   Schuiling M, Zuidhof AB, Zaagsma J, Meurs H (1999) Role of tachykinin NK1 and NK2 receptors in allergen-induced early and late asthmatic reactions, airway hyperresponsiveness, and airway inflammation in conscious, unrestrained guinea pigs. *Clin Exp Allergy* 29 S2: 48–52

74   Schuiling M, Zuidhof AB, Meurs H, Zaagsma J (1999) Role of tachykinin NK2-receptor activation in the allergen-induced late asthmatic reaction, airway hyperreactivity and airway inflammatory cell influx in conscious, unrestrained guinea-pigs. *Br J Pharmacol* 127: 1030–1038

75 Mukaiyama O, Morimoto K, Nosaka E, Takahashi S, Yamashita M (2004) Involvement of enhanced neurokinin NK3 receptor expression in the severe asthma guinea pig model. *Eur J Pharmacol* 498: 287–294

76 Mukaiyama O, Morimoto K, Nosaka E, Takahashi S, Yamashita M (2004) Greater involvement of neurokinins found in Guinea pig models of severe asthma compared with mild asthma. *Int Arch Allergy Immunol* 134: 263–272

77 Meurs H, McKay S, Maarsingh H, Hamer MA, Macic L, Molendijk N, Zaagsma J (2002) Increased arginase activity underlies allergen-induced deficiency of cNOS-derived nitric oxide and airway hyperresponsiveness. *Br J Pharmacol* 136: 391–398

78 Bautsch W, Hoymann HG, Zhang Q, Meier-Wiedenbach I, Raschke U, Ames RS, Sohns B, Flemme N, Meyer zu Vilsendorf A, Grove M et al (2002) Cutting edge: guinea pigs with a natural C3a-receptor defect exhibit decreased bronchoconstriction in allergic airway disease: evidence for an involvement of the C3a anaphylatoxin in the pathogenesis of asthma. *J Immunol* 165: 5401–5405

79 Morse B, Sypek JP, Donaldson DD, Haley KJ, Lilly CM (2002) Effects of IL-13 on airway responses in the guinea pig. *Am J Physiol Lung Cell Mol Physiol* 282: L44–49

80 Pauluhn J, Woolhiser MR, Bloemen L (2005) Repeated inhalation challenge with diphenylmethane-4,4'-diisocyanate in brown Norway rats leads to a time-related increase of neutrophils in bronchoalveolar lavage after topical induction. *Inhal Toxicol* 17: 67–78

81 Zhang XD, Fedan JS, Lewis DM, Siegel PD (2004) Asthmalike biphasic airway responses in Brown Norway rats sensitized by dermal exposure to dry trimellitic anhydride powder. *J Allergy Clin Immunol* 113: 320–326

82 Dong W, Selgrade MK, Gilmour MI (2003) Systemic administration of *Bordetella pertussis* enhances pulmonary sensitization to house dust mite in juvenile rats. *Toxicol Sci* 72: 113–121

83 Djuric VJ, Cox G, Overstreet DH, Smith L, Dragomir A, Steiner M (1998) Genetically transmitted cholinergic hyperresponsiveness predisposes to experimental asthma. *Brain Behav Immun* 12: 272–284

84 Leung SY, Eynott P, Nath P, Chung KF (2005) Effects of ciclesonide and fluticasone propionate on allergen-induced airway inflammation and remodeling features. *J Allergy Clin Immunol* 115: 989–996

85 Eynott PR, Nath P, Leung SY, Adcock IM, Bennett BL, Chung KF (2003) Allergen-induced inflammation and airway epithelial and smooth muscle cell proliferation: role of Jun N-terminal kinase. *Br J Pharmacol* 140: 1373–1380

86 Salmon M, Walsh DA, Koto H, Barnes PJ, Chung KF (1999) Repeated allergen exposure of sensitized Brown-Norway rats induces airway cell DNA synthesis and remodelling. *Eur Respir J* 14: 633–641

87 Xu KF, Vlahos R, Messina A, Bamford TL, Bertram JF, Stewart AG (2002) Antigen-induced airway inflammation in the Brown Norway rat results in airway smooth muscle hyperplasia. *J Appl Physiol* 93: 1833–1840

88   Vanacker NJ, Palmans E, Pauwels RA, Kips JC (2002) Fluticasone inhibits the progression of allergen-induced structural airway changes. *Clin Exp Allergy* 32: 914–920

89   Trifilieff A, Wyss D, Walker C, Mazzoni L, Hersperger R (2002) Pharmacological profile of a novel phosphodiesterase 4 inhibitor, 4-(8-benzo[1,2,5]oxadiazol-5-yl-[1,7]naphthyridin-6-yl)-benzoic acid (NVP-ABE171), a 1,7-naphthyridine derivative, with anti-inflammatory activities. *J Pharmacol Exp Ther* 301: 241–248

90   Aoki M, Fukunaga M, Kitagawa M, Hayashi K, Morokata T, Ishikawa G, Kubo S, Yamada T (2000) Effect of a novel anti-inflammatory compound, YM976, on antigen-induced eosinophil infiltration into the lungs in rats, mice, and ferrets. *J Pharmacol Exp Ther* 295: 1149–1155

91   Kurucz I, Nemeth K, Meszaros S, Torok K, Nagy Z, Zubovics Z, Horvath K, Bodor N (2004) Anti-inflammatory effect and soft properties of etiprednol dicloacetate (BNP-166), a new, anti-asthmatic steroid. *Pharmazie* 59: 412–416

92   Hersperger R, Buchheit KH, Cammisuli S, Enz A, Lohse O, Ponelle M, Schuler W, Schweitzer A, Walker C, Zehender H et al (2004) A locally active antiinflammatory macrolide (MLD987) for inhalation therapy of asthma. *J Med Chem* 47: 4950–4957

93   del Pozo V, Rojo M, Rubio ML, Cortegano I, Cardaba B, Gallardo S, Ortega M, Civantos E, Lopez E, Martin-Mosquero C et al (2002) Gene therapy with galectin-3 inhibits bronchial obstruction and inflammation in antigen-challenged rats through interleukin-5 gene downregulation. *Am J Respir Crit Care Med* 166: 732–737

94   Stenton GR, Ulanova M, Dery RE, Merani S, Kim MK, Gilchrist M, Puttagunta L, Musat-Marcu S, James D, Schreiber AD et al (2002) Inhibition of allergic inflammation in the airways using aerosolized antisense to Syk kinase. *J Immunol* 169: 1028–1036

95   Ziegelbauer K, Gantner F, Lukacs NW, Berlin A, Fuchikami K, Niki T, Sakai K, Inbe H, Takeshita K, Ishimori M et al (2005) A selective novel low-molecular-weight inhibitor of IkappaB kinase-beta (IKK-beta) prevents pulmonary inflammation and shows broad anti-inflammatory activity. *Br J Pharmacol* 145: 178–192

96   Eynott PR, Xu L, Bennett BL, Noble A, Leung SY, Nath P, Groneberg DA, Adcock IM, Chung KF (2004) Effect of an inhibitor of Jun N-terminal protein kinase, SP600125, in single allergen challenge in sensitized rats. *Immunology* 112: 446–453

97   Morokata T, Suzuki K, Ida K, Tsuchiyama H, Ishikawa J, Yamada T (2004) Effect of a novel interleukin-5 receptor antagonist, YM-90709 (2,3-dimethoxy-6,6-dimethyl-5,6-dihydrobenzo[7,8]indolizino[2,3-b]quinoxaline), on antigen-induced airway inflammation in BN rats. *Int Immunopharmacol* 4: 873–883

98   Huang TJ, Eynott P, Salmon M, Nicklin PL, Chung KF (2002) Effect of topical immunomodulators on acute allergic inflammation and bronchial hyperresponsiveness in sensitised rats. *Eur J Pharmacol* 437: 187–194

99   Eynott PR, Salmon M, Huang TJ, Oates T, Nicklin PL, Chung KF (2003) Effects of cyclosporin A and a rapamycin derivative (SAR943) on chronic allergic inflammation in sensitized rats. *Immunology* 109: 461–467

100 Matsumoto K, Hayakawa H, Ide K, Suda T, Chida K, Hashimoto H, Sato A, Nakamu-

ra H (2002) Effects of suplatast tosilate on cytokine profile of bronchoalveolar cells in allergic inflammation of the lung. *Respirology* 7: 201–207

101 Fozard JR, Ellis KM, Villela Dantas MF, Tigani B, Mazzoni L (2002) Effects of CGS 21680, a selective adenosine A2A receptor agonist, on allergic airways inflammation in the rat. *Eur J Pharmacol* 438: 183–188

102 Huang TJ, Adcock IM, Chung KF (2001) A novel transcription factor inhibitor, SP100030, inhibits cytokine gene expression, but not airway eosinophilia or hyperresponsiveness in sensitized and allergen-exposed rat. *Br J Pharmacol* 134: 1029–1036

103 Nonaka T, Mitsuhashi H, Takahashi K, Sugiyama H, Kishimoto T (2000) Effect of TEI-9874, an inhibitor of immunoglobulin E production, on allergen-induced asthmatic model in rats. *Eur J Pharmacol* 402: 287–295

104 Elliott PJ, Pien CS, McCormack TA, Chapman ID, Adams J (1999) Proteasome inhibition: A novel mechanism to combat asthma. *J Allergy Clin Immunol* 104: 294–300

105 Eber E, Uhlig T, McMenamin C, Sly PD (1998) Leflunomide, a novel immunomodulating agent, prevents the development of allergic sensitization in an animal model of allergic asthma. *Clin Exp Allergy* 28: 376–384

106 Hazarika S, Van Scott MR, Lust RM (2004) Myocardial ischemia-reperfusion injury is enhanced in a model of systemic allergy and asthma. *Am J Physiol Heart Circ Physiol* 286: H1720–1725

107 Gallelli L, D'Agostino B, Marrocco G, De Rosa G, Filippelli W, Rossi F, Advenier C (2003) Role of tachykinins in the bronchoconstriction induced by HCl intraesophageal instillation in the rabbit. *Life Sci* 72: 1135–1142

108 Hogman M, Hjoberg J, Almirall J, Hedenstierna G (1999) Both inhaled histamine and hypertonic saline increase airway reactivity in non-sensitised rabbits. *Respiration* 66: 349–354

109 Grunstein MM, Veler H, Shan X, Larson J, Grunstein JS, Chuang S (2005) Proasthmatic effects and mechanisms of action of the dust mite allergen, Der p 1, in airway smooth muscle. *J Allergy Clin Immunol* 116: 94–101

110 Grunstein MM, Hakonarson H, Maskeri N, Chuang S (2000) Autocrine cytokine signaling mediates effects of rhinovirus on airway responsiveness. *Am J Physiol Lung Cell Mol Physiol* 278: L1146–1153

111 Whelan R, Kim C, Chen M, Leiter J, Grunstein MM, Hakonarson H (2004) Role and regulation of interleukin-1 molecules in pro-asthmatic sensitised airway smooth muscle. *Eur Respir J* 24: 559–567

112 Gascoigne MH, Holland K, Page CP, Shock A, Robinson M, Foulkes R, Gozzard N (2003) The effect of anti-integrin monoclonal antibodies on antigen-induced pulmonary inflammation in allergic rabbits. *Pulm Pharmacol Ther* 16: 279–285

113 Obiefuna PC, Batra VK, Nadeem A, Borron P, Wilson CN, Mustafa SJ (2005) A novel A1 adenosine receptor antagonist, L-97-1 [3-[2-(4-aminophenyl)-ethyl]-8-benzyl-7-{2-ethyl-(2-hydroxy-ethyl)-amino]-ethyl}-1-propyl-3,7-dihydro-purine-2,6-dione], reduces allergic responses to house dust mite in an allergic rabbit model of asthma. *J Pharmacol Exp Ther* 315: 329–336

114 Woisin FE, Matsumoto T, Douglas GJ, Paul W, Whalley ET, Page CP (2000) Effect of antagonists for NK(2)and B(2) receptors on antigen-induced airway responses in allergic rabbits. *Pulm Pharmacol Ther* 13: 13–23

115 Assa'ad AH, Ballard ET, Sebastian KD, Loven DP, Boivin GP, Lierl MB (1998) Effect of superoxide dismutase on a rabbit model of chronic allergic asthma. *Ann Allergy Asthma Immunol* 80: 215–224

116 Tiniakov RL, Tiniakova OP, McLeod RL, Hey JA, Yeates DB (2003) Canine model of nasal congestion and allergic rhinitis. *J Appl Physiol* 94: 1821–1828

117 Skorohod N, Yeates DB (2005) Superoxide dismutase failed to attenuate allergen-induced nasal congestion in ragweed-sensitized dogs. *J Appl Physiol* 98: 1478–1486

118 House A, Celly C, Young S, Kreutner W, Chapman RW (2001) Bronchoconstrictor reactivity to NKA in allergic dogs: a comparison to histamine and methacholine. *Pulm Pharmacol Ther* 14: 135–140

119 Redman TK, Rudolph K, Barr EB, Bowen LE, Muggenburg BA, Bice DE (2002) Pulmonary immunity to ragweed in a Beagle dog model of allergic asthma. *Exp Lung Res* 27: 433–451

120 Barrett EG, Rudolph K, Bowen LE, Bice DE (2003) Parental allergic status influences the risk of developing allergic sensitization and an asthmatic-like phenotype in canine offspring. *Immunology* 110: 493–500

121 Freed AN, Davis MS (1999) Hyperventilation with dry air increases airway surface fluid osmolality in canine peripheral airways. *Am J Respir Crit Care Med* 159: 1101–1107

122 Suzuki R, Freed AN (2000) Heparin inhibits eicosanoid metabolism and hyperventilation-induced bronchoconstriction in dogs. *Am J Respir Crit Care Med* 161: 1850–1854

123 Davis MS, Schofield B, Freed AN (2003) Repeated peripheral airway hyperpnea causes inflammation and remodeling in dogs. *Med Sci Sports Exerc* 35: 608–616

124 Abraham WM, Ahmed A, Sabater JR, Lauredo IT, Botvinnikova Y, Bjercke RJ, Hu X, Revelle BM, Kogan TP, Scott IL et al (1999) Selectin blockade prevents antigen-induced late bronchial responses and airway hyperresponsiveness in allergic sheep. *Am J Respir Crit Care Med* 159: 1205–1214

125 Abraham WM, Gill A, Ahmed A, Sielczak MW, Lauredo T, Botinnikova Y, Lin KC, Pepinsky B, Leone D, Lobb RR, Adams SP (2000) A small molecule, tight binding inhibitor of the integrin alpha4 beta1 blocks antigen-induced airway responses and inflammation inexperimental asthma in sheep. *Am J Respir Care Med* 162: 603–611

126 Abraham WM, Bourdelais AJ, Sabater JR, Ahmed A, Lee TA, Serebriakov I, Baden DG (2005) Airway responses to aerosolized brevetoxins in an animal model of asthma. *Am J Respir Crit Care Med* 171: 26–34

127 O'Riordan TG, Mao Y, Otero R, Lopez J, Sabater JR, Abraham WM (1998) Budesonide affects allergic mucociliary dysfunction. *J Appl Physiol* 85: 1086–1091

128 Costanza MJ, Yabut SC, Almond HR, Andrade-Gordon P, Corcoran TW, de Garavilla L, Kauffman JA, Abraham WM, Recacha R, Chattopadhyay D, Maryanoff BE (2003) Potent, small molecule inhibitors of human mast cell tryptase. Antiasthmatic action of a

dipeptide-based transition state analogue containing a benzothiazole ketone. *J Med Chem* 46: 3865–3876

129 Krishna MT, Chauhan A, Little L, Sampson K, Hawksworth R, Mant T, Djukanovic R, Lee T, Holgate S (2001) Inhibition of mast cell tryptase by inhaled APC 366 attenuates allergen-induced late-phase airway obstruction in asthma. *J Allergy Clin Immunol* 107: 1039–1045

130 Ahmed T, Ungo J, Zhou M, Campo C (2000) Inhibition of allergic late airway responses by inhaled heparin-derived oligosaccharides. *J Appl Physiol* 88: 1721–1729

131 Scuri M, Allegra L, Abraham WM (1998) The effects of multiple dosing with zileuton on antigen-induced responses in sheep. *Pulm Pharmacol Ther* 11: 277–280

132 Sabater JR, Wanner A, Abraham WM (2002) Montelukast prevents antigen-induced mucociliary dysfunction in sheep. *Am J Respir Crit Care Med* 166: 1457–1460

133 Lauredo IT, Forteza RM, Botvinnikova Y, Abraham WM (2004) Leukocytic cell sources of airway tissue kallikrein. *Am J Physiol Lung Cell Mol Physiol* 286: L734–740

134 Bischoff R, Snibson K, Shaw R, Meeusen EN (2003) Induction of allergic inflammation in the lungs of sensitized sheep after local challenge with house dust mite. *Clin Exp Allergy* 33: 367–375

135 Snibson KJ, Bischof RJ, Slocombe RF, Meeusen EN (2005) Airway remodelling and inflammation in sheep lungs after chronic airway challenge with house dust mite. *Clin Exp Allergy* 35: 146–152

136 Singh J, van Vlijmen H, Liao Y, Lee WC, Cornebise M, Harris M, Shu I, Gill A, Cuervo JH, Abraham WM, Adams SP (2002) Identification of potent and novel alpha4 beta1 antagonists using in silico screening. *J Med Chem* 45: 2988–2993

137 Abraham WM, Ahmed A, Serebriakov I, Carmillo AN, Ferrant J, de Fougerolles AR, Garber EA, Gotwals PJ, Koteliansky VE, Taylor F, Lobb RR (2003) A monoclonal antibody to alpha1 beta1 blocks antigen-induced airway responses in sheep. *Am J Respir Crit Care Med* 169: 97–104

138 Rosen SD, Tsay D Singer MS, Hemmerich S, Abraham WM (2005) Therapeutic targeting of endothelial ligands for L-selectin (PNAd) in a sheep model of asthma. *Am J Pathol* 166: 935–944

139 Kasserra CE, Harris P, Stenton GR, Abraham W, Langlands JM (2004) IPL576,092, a novel anti-inflammatory compound, inhibits leukocyte infiltration and changes in lung function in response to allergen challenge. *Pulm Pharmacol Ther* 17: 309–318

140 Freyne EJ, Lacrampe JF, Deroose F, Boeckx GM, Willems M, Embrechts W, Coesemans E, Willems JJ, Fortin JM, Ligney Y et al (2005) Synthesis and biological evaluation of 1,2,4-triazinylphenylalkylthiazolecarboxylic acid esters as cytokine-inhibiting antidrugs with strong bronchodilating effects in an animal model of asthma. *J Med Chem* 48: 2167–2175

141 Fujimoto K, Tsunoda T, Koizumi T, Kubo K (2002) Effects of an eosinophil allergic sheep. *Lung* 180: 161–172

142 Black KR, Suki B, Madwed JB, Jackson AC (2001) Airway resistance and tissue elas-

tance from input or transfer impedance in bronchoconstricted monkeys. *J Appl Physiol* 90: 571–578

143 Schelegle ES, Gershwin LJ, Miller LA, Fanucchi MV, Van Winkle LS, Gerriets JP, Walby WF, Omlor AM, Buckpitt AR, Tarkington BK et al (2001) Allergic asthma induced in rhesus monkeys by house dust mite (*Dermatophagoides farinae*). *Am J Pathol* 158: 333–341

144 Coffman RL, Hessel EM (2005) Nonhuman primate models of asthma. *J Exp Med* 201: 1875–1879

145 Van Scott MR, Hooker JL, Ehrmann D, Shibata Y Kukoly C, Salleng K, Westergaard G, Sandrasagra A Nyce J (2004) Dust mite-induced asthma in cynomolgus monkeys. *J Appl Physiol* 96: 1433–1444

146 Van Scott MR, Aycock D, Cozzi E, Salleng K, Stallings HW (2005) Separation of bronchoconstriction from increased ventilatory drive in a nonhuman primate model of chronic allergic asthma. *J Appl Physiol* 99: 2080–2086

147 Hart TK, Cook RM, Zia-Amirhosseini P, Minthorn E, Sellers TS, Maleeff BE, Eustis S, Schwartz LW, Tsui P, Applebaum ER et al (2001) Preclinical efficacy and safety of mepolizumab (SB-240563), a humanized monoclonal antibody to IL-5, in cynmolgus monkeys. *J Allergy Clin Immunol* 108: 250–257

148 Fanucchi MV, Schelegle ES, Baker GL, Evans MJ, NcDonald RJ, Gershwin LJ, Raz E, Hyde DM, Plopper CG, Miller LA (2004) Immunostimulatory oligonucleotides attenuate airways remodeling in allergic monkeys. *Am J Resp Crit Care Med* 170: 1153–1157

149 Buchheit KH, Manley PW, Quast U, Russ U, Mazzoni L, Fozard JR (2002) KCO912: a potent and selective opener of ATP-dependent potassium (K(ATP)) channels which suppresses airways hyperreactivity at doses devoid of cardiovascular effects. *Naunyn Schmiedebergs Arch Pharmacol* 365: 220–230

150 Billah MM, Cooper N, Minnicozzi M, Warneck J, Wang P, Hey JA, Kreutner W, Rizzo CA, Smith SR, Young S et al (2002) Pharmacology of N-(3,5-dichloro-1-oxido-4-pyridinyl)-8-methoxy-2-(trifluoromethyl)-5-quinoline carboxamide (SCH 351591), a novel, orally active phosphodiesterase 4 inhibitor. *J Pharmacol Exp Ther* 302: 127–137

In vivo modeling systems for chronic obstructive pulmonary disease

Christopher S. Stevenson[1] and David C. Underwood[2]

[1]Respiratory Disease Area, Novartis Institutes for Biomedical Research, Novartis Horsham, West Research Centre, Wimblehurst Road, Horsham Sussex, RH12 5AB, United Kingdom; [2]Center for Excellence in Drug Discovery, Respiratory Diseases, GlaxoSmithKline Pharmaceuticals, 709 Swedeland Road, King of Prussia, PA 19406, USA

Introduction

This chapter is designed to act as a reference for researchers in the field of chronic obstructive pulmonary disease (COPD). Provided are descriptions of the diverse manifestations of the disease and the *in vivo* models that are commonly used to mirror them. COPD is a smoking-related disorder that is a major cause of morbidity and mortality throughout the world. It comprises a group of lung conditions commonly described clinically as chronic bronchitis, small airways disease, and emphysema. While no single animal model replicates the degree of lung destruction observed in the human disease condition, there are models that can mimic many of the same types of pathologies using disease-relevant agents. Several modeling systems have been developed over the years to mimic aspects of COPD, and it is therefore beyond the scope of this chapter to be a comprehensive review of all them; however, where possible, tables citing the relevant publications on these diverse models have been included for the readers' reference. Instead, this chapter focuses on the many recent advances in two classic modeling systems which use elastase or cigarette smoke to replicate the destructive pathologies associated with COPD. A short review of transgenic mouse models that resemble pathological aspects of COPD is also included.

Chronic obstructive pulmonary disease

COPD is not one disease, but a disease spectrum encompassing complications from reductions in airflow due to structural changes to the lung. The disease is located primarily in the airways, but there are some manifestations in the pulmonary vasculature and in the skeletal muscle. In the central airways there is mucus gland

In Vivo Models of Inflammation, Vol. II, edited by Christopher S. Stevenson, Lisa A. Marshall and Douglas W. Morgan

enlargement, goblet cell metaplasia, epithelial thickening and mucus hypersecretion (clinically characterized as chronic bronchitis). In the small airways there is subepithelial fibrosis and inflammatory exudates contributing to fixed airway obstruction (characterized as small airways disease). At the bronchiole-alveolar junction, the respiratory bronchioles are dilated and the alveolar walls are destroyed, resulting in increased lung compliance (described as emphysema). The occurrence of any of these pathologies can vary considerably between patients. These changes occur as focal lesions in the lung and their contribution to the loss in lung function significantly overlap. In addition, like asthma, the airflow obstruction may be accompanied by airway hyperresponsiveness, cough, and enhanced mucus production. However, unlike asthma, the disease-related remodeling of the airways in COPD is more prominent, and the airflow reductions, generally assessed by measurement of upper and central airway flow, are only partially reversible by bronchodilators [1].

The inflammation associated with COPD is elevated compared to that observed in healthy smokers, and is currently not treatable even by conventional anti-inflammatory therapies, i.e., glucocorticoids [2]. It is also believed to be a key driving force for the resulting pathological changes that result in the compromised lung function. There are increased numbers of neutrophils, macrophages and lymphocytes in various parts of the lung. Neutrophils and macrophages release inflammatory cytokines, proteases, and additional oxidants that are thought to contribute to the lung pathologies associated with COPD, especially early in the disease process [3–5]. B cells, CD4+, and CD8+ T cell numbers are also increased in the lung, although only the amount of B cells and CD8+ T cells inversely correlate to lung function [6–8]. The elevation in the lymphocyte numbers is also associated with an increase in bronchial associated lymphoid tissue in the adventitia of the small airways in the later stages of the disease [8]. These later changes in the adaptive immune response may be indicative of the bacterial colonization of the normally sterile lung and greater incidences of exacerbation (triggered by viral, bacterial, or environmental insults) that worsen with disease progression [9]. It may also signal that there may be an autoimmune component in the later stages of the disease process [10–12].

COPD is prevalent in about 20 million men and women in the United States, mostly over 40 years of age. The mortality is about 20/100 000, making it the fourth leading cause of death in the United States, and it is expected to be the third leading cause of death worldwide by 2020 [13]. Over 90% of COPD cases in Western societies are due to years of heavy cigarette smoking. Smoking is also a common etiological factor in rapidly developing nations like China and India; however, occupational and environmental exposure are also major causes in these countries. Interestingly, only 15–50% of smokers will develop COPD [14, 15]. This suggests that there are likely to be a number of genetic risk factors associated with developing COPD, but these remain largely unknown. There is, however, one well-documented genetic risk factor associated with an early on-set and accelerated form

Table 1 - Representative animal models of mucus hypersecretion

Agent	Species
Sulfur dioxide	Rat, hamster, dog
Nitrogen dioxide	Rat, hamster
Elastase	Rat, hamster
Cigarette smoke	Rat, mouse, guinea pig, dog, lamb
Ozone	Rat, monkey
Lipopolysaccharide (endotoxin)	Rat, mouse, guinea pig
Metabisulfite	Rat, guinea pig
Elastase	Rat, hamster
Transgenic mouse strain	IL-13$^{+/+}$
	IFN-$\gamma^{+/+}$
	Human IL-1$\beta^{+/+}$

Reviewed in [126, 127]

of COPD. Individuals deficient in α1-antitrypsin, the endogenous inhibitor of neutrophil elastase (NE), develop panacinar emphysema due to the resulting proteolytic imbalance in the lung [16, 17]. While the disease generally manifests itself in smokers, individuals with a severe deficiency can also develop emphysema as nonsmokers.

Because of the general irreversibility of airflow obstruction with present therapy, especially in severe fibrosis or emphysema, treatment approaches have been directed towards optimizing the ventilation-perfusion relationship (oxygen therapy and surgical lung reduction) and reducing progress of the disease (cessation of exposure). Thus, quality of life assessment is a crucial marker, and presently considered unapproachable through animal modeling. Currently, there are no drugs available that can reverse or halt the progression of COPD. A major focus has been to identify anti-inflammatory therapies that can dampen the enhanced inflammation occurring in the lungs of these patients. Therefore, much of the work of modeling COPD in animals has focused on mimicking the inflammatory changes and the resulting pathologies that cause the loss in lung function associated with the disease. As mentioned previously, there have been many models developed to mimic aspects of the disease including mucus hypersecretion (Tab. 1), emphysema (Tabs 2 and 3), and airway fibrosis (Tab. 4). In this chapter, the focus will remain on those rodent models that are commonly used to investigate disease mechanisms and profile candidate compounds that target the inflammation and resulting pathologies (particularly emphysema) associated with COPD.

Table 2 - Representative animal models of emphysema

Agent	Species
Elastase	Rat, hamster, guinea pig, rabbit
Cigarette smoke	Guinea pig, mouse, dog
Endotoxin	Mouse, hamster, dog
Nitrogen dioxide (NO$_2$)	Mice, rats
Cadmium	Rats, guinea pigs
Occupational mineral (coal, quartz, silica dust)	Rat, hamster, guinea pig
Pharmacological	
VEGF inhibition	Mice
methylprednisolone	Mice
Autoimmune (xenogeneic endothelial delivered)	Mouse

Reviewed in [20, 21, 126, 128, 129]

Elastase-induced lung inflammation and emphysema

To date, the most important contribution animal modeling has made to our understanding of COPD pathogenesis was the discovery that instillation of the plant proteinase papain to rat lung induces emphysema [18]. This finding, along with the clinical observation that patients deficient in α1-antitrypsin developed early onset emphysema, shaped the proteinase-antiproteinase imbalance hypothesis that explains how the emphysematous damage associated with COPD may develop [16, 17]. Emphysema is characterized by abnormal permanent enlargement of airspaces distal to the terminal bronchiole associated with the loss of extracellular matrix, particularly elastin, in the lung. This leads to a loss of blood-gas exchange surface area, affects small airway patency, and causes lung hyperinflation, first during exercise and, as the disease progresses, also at rest. Although certain diagnosis is made only through histological examination of inflated, fixed whole lung, demonstrable airway obstruction may be detected due to severely diminished elastic recoil of affected tissues, which allows the airways to collapse during expiration [19]. Centriacinar emphysema describes dilation of the respiratory bronchioles, where the bronchiole joins the alveolar duct. This pattern is consistent with the deposition of inhaled toxicants in lungs, usually localizing in the upper lobes, particularly in the apical segments, and is the most common form in smokers [20–22]. Panacinar emphysema is more commonly found in patients with α1-antitrypsin deficiency and involves the uniform enlargement of airspaces distal to the terminal bronchioles [20]. These lesions are more common in the lower zones and anterior margins of the lung [22]. Regardless of its classification or distribution, airspace enlargement

Table 3 - Representative genetic models of emphysema

Type of mutant	Strain
Naturally occurring genetic mutant mouse strains with emphysema	Tight-skinned (Tsk)
	Blotchy
	Pallid (PA)
	Beige (Bg)
	Osteopetrotic (M-CSF$^{-/-}$)
Spontaneous emphysema in mice with gene deletions	Surfactant D$^{-/-}$
	Integrin $\alpha\nu\beta6^{-/-}$
	Tissue inhibitor of metalloproteinases-(TIMP-)3$^{-/-}$
Inducible transgenic mouse strain	IL-13$^{+/+}$
	IFN-$\gamma^{+/+}$
	Human IL-1$\beta^{+/+}$
	TNF-$\alpha^{+/+}$

Reviewed in [128]

Table 4 - Representative animal models of lung fibrosis

Agent	Species and/or strain
Bleomycin	Mice, rats, hamsters, pheasants, dogs, sheep
Cadmium chloride	Mice, rat
Cigarette smoke	Rat
Irradiation	Mice, rats, hamsters, dogs, sheep
Lipopolysaccharide	Mice
Inorganic particles	Mice, rats, hamsters, sheep
Transgenic mouse strain	IL-13$^{+/+}$
	IFN-$\gamma^{+/+}$
	Human IL-1$\beta^{+/+}$
	TNF-$\alpha^{+/+}$
Vanadium	Rat
3-Methylfuran	Rat
Adenoviral gene transfer	
TGF-β	Rat
IL-1β	Mice
Nitric acid	Dog

Reviewed in [127, 130]

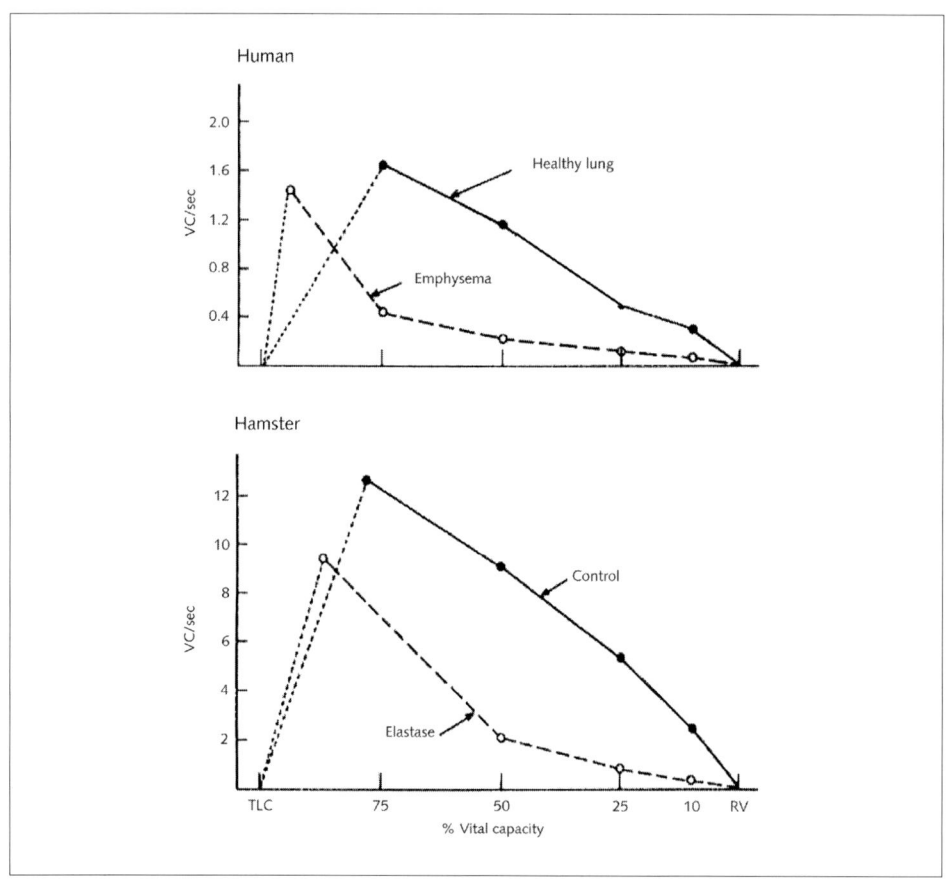

Figure 1
The upper graph demonstrates the differences in lung vital capacity that are evident with emphysema (adapted from [123]), and a similar depiction (lower graph) of data in the emphysematous state produced by intratracheal instillation of human neutrophil elastase in the hamster (adapted from [124]).

resulting from destruction of alveolar walls constitutes the morphological hallmarks of this disease.

Generally, the model involves a single intratracheal instillation of elastase or papain into the lung (Tab. 2; [20]). Administration of elastase induces a transient acute inflammatory response, followed by airspace enlargement that resembles the panacinar form of emphysema [20]. Figure 1 demonstrates the substantial change in lung capacity that occurs in human emphysema (adapted from [23]), and a similar depiction of data from elastase-induced emphysema in the hamster (adapted from

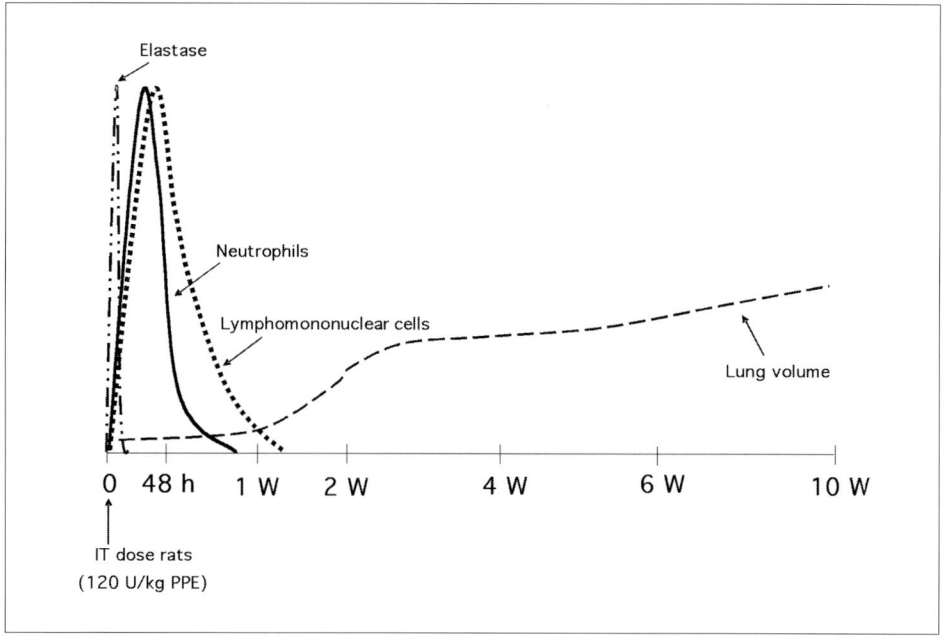

Figure 2
The time course of events following elastase administration in the rat lung (adapted from
[125]). Neutrophil elastase activity is present in the BAL for a very short period, followed by
a transient inflammatory response (resolving within a week) and airspace enlargement.

[24]). Papain or purified preparations of bovine, porcine, or human pancreatic elastase or NE have been most commonly used to produce emphysematous lesions in animals [20]. Because of their relatively greater sensitivity to elastolytic enzymes, hamsters had initially been used more often to model emphysema. In more recent years, both rats and mice have been used to investigate the mechanisms involved in elastase-mediated lung destruction as there are more molecular tools available for these species. A single instillation of elastase into the upper airways results in greater elastolytic activity in the bronchoalveolar lavage fluid (BALF) for up 24 h. Airspace enlargement begins immediately, followed by an acute alveolitis characterized by an early neutrophilic and lymphomononuclear invasion that generally resolves within a week (Fig. 2; adapted from [25]). It is believed that degradation of lung elastin and collagen generates chemoattractant fragments that helps drive this inflammatory response [26–28]. Further, because the elastolytic activity resolves very quickly, the continued destruction of the lung is believed to be due in part to inflammatory proteinases [25, 29]. This further degradation causes secretory cell metaplasia, rupture

of the alveolar epithelium, pulmonary edema, hemorrhage and airspace enlargement, which results in decreased lung function [25, 30, 31]. The rapid synthesis of new elastin and collagen cannot repair the normal lung architecture, resulting in a permanent distortion and derangement of alveolar structures [31].

This model has been used to investigate mechanisms that can provide insights into the development as well as the reversal of emphysematous destruction in the lung. One of the most important investigations into the reversal emphysematous damage came from the studies conducted by Massaro and Massaro [32]. They and others identified that vitamin A, retinoid binding proteins, and retinoic acid receptors were increased in developing lungs and decreased in lungs where development was inhibited by dexamethasone [34–37]. These findings led to the discovery that all-trans-retinoic acid could increase the number of alveoli in a normal rat lung [33]. Following this, the Massaros confirmed that all-trans-retinoic acid could also reverse the emphysema induced by elastase administration [32]. However, this finding remains controversial as a number of other groups have not been able to demonstrate this effect in other species, and there has been no positive results reported from ongoing clinical studies [38–41].

Additional pharmacological studies have shown that this model may have more in common with the clinical disease than previously thought. Birrell and colleagues [25] have shown that the inflammation and physiological changes induced by elastase are not sensitive to prophylactic or therapeutic administration of steroids, a hallmark of COPD. They suggested the reason for the steroid insensitivity was the inflammation was independent of transcription factor nuclear factor-κB (NF-κB) activation. Subsequently, they demonstrated that an inhibitor of IκB kinase, a key kinase involved in classical NF-κB pathway activation, has no impact on the airway inflammation. [41]. Additional studies using knockout mice have shown that mice deficient in TNF-α or IL-1β receptors were partially protected from elastase-induced lung destruction [42]. When both receptors were knockout, airspace sizes were comparable with that of the control. This suggests that the inflammation induced by elastase is equally or more important to the development of emphysema in this model than the parenchymal destruction caused by elastase alone. Other groups have demonstrated that oxidant stress may also contribute to the damage caused by elastase. Treatment with N-acetylcysteine and adenovirus transfection of heme oxygenase-1 have been shown to be protective in this model [43–45] . In addition, mice deficient in nuclear factor erythroid2-related factor 2 (Nrf2), the transcription factor responsible for activating the antioxidant response element, have been shown to develop enhanced inflammation and emphysema after elastase administration [46]. Oxidants may be produced by the infiltrating inflammatory cells and contribute to lung degradation by activating inflammatory signaling pathways [47, 48] and inactivating endogenous proteinase inhibitors such α1-antitrypsin [48, 49], leading to greater proteolytic activity in the lung. Additional findings point to the importance of the matrix elements in the lung. Cantor and colleagues [50] were able to demonstrate that administration of hyaluronidase could prevent elastase-induced emphysema in hamsters, suggesting that lung elastin could potentially be "shielded" from elastases.

The most important benefits of this model are that it generates a relevant lung pathology in a relatively short period of time and allows investigators to study the mechanisms involved in lung repair. It also allows one to look at the effect of candidate compounds using both prophylactic and therapeutic dosing regimens. The limitations of this model is that the lung destruction in this model results from a transient increase in a single mediator. This is unlike the clinical situation, which results from years of repetitive insults due to chronic cigarette smoking, which contains particulate matter, free radicals, reactive oxygen species, and over 4000 chemicals. How well elastase administration reflects the complexity of these signals involved in disease pathogenesis is unknown. Further, the inflammatory changes are only transient, unlike COPD where the inflammation subsists years after smoking cessation. In addition, while some of the changes that occur appear to be similar to those observed clinically (e.g., emphysema and goblet cell metaplasia), it does not replicate all of the changes (e.g., small airway remodeling). Therefore, while this model can be used to model mechanistic aspects of the disease, its ability to predict the potential efficacy of the candidate compounds is still unclear.

Cigarette smoke-induced inflammation and lung destruction

Cigarette smoke is another agent used to model the changes associated with COPD. Its obvious advantages include that it is the most clinically relevant stimulus and, therefore, may be more likely to induce the pathways that lead to disease. The inflammatory changes in these models appear to be consistent to that observed in man and result from repetitive exposures to cigarette smoke – another advantage over other models (e.g., elastase). In addition, cigarette smoke has also been able to induce most of the relevant COPD-related lung pathologies (chronic low-grade inflammation, mucus hypersecretion, airway fibrosis, emphysema, and pulmonary hypertension). One of the limitations of this model is that there appears to be significant differences in the response and sensitivity to smoke inhalation between some species as well as strains [51–54]. The majority of the work in this area is done using mice, rats, and guinea pigs; therefore, the descriptions of the changes due to smoke inhalation given here are a general summary of findings from these species. Other disadvantages include access to smoke exposure equipment is not available to many laboratories, animals need to be exposed on a daily basis, and it generally takes a long time to generate relatively mild pathologies (approximately 3 months minimum); thus, these models are labor intensive.

There are a variety of machines that can be used for generating cigarette smoke and a number of different types of exposure systems used for these experiments. To go through them all is beyond the scope of this review, but the two most common methods for delivering cigarette smoke to small rodents are either "nose-only" or "whole-body" exposure systems. While the methods of delivery may be different, the results generated from both systems have been relatively comparable [55–59]. Researchers using both systems titrate the amount smoke delivered to animals based on the amount of total particulate matter entering the exposure chambers, the carboxyhemaglobin levels in the blood of the animals after exposure, CO levels in the chambers, or cotinine levels in serum. Establishing such guidelines allow individuals to check that animals are exposed uniformly and develop a standard smoke-exposure protocol, which includes a set volume of smoke delivered over a specific period of time generated from a certain number of cigarettes.

Acute smoking models

Because of the time it takes to run the more chronic models that generate COPD-like pathologies, acute ($\leq$ 3 days) and sub-chronic models (<4 weeks) have been developed to evaluate the dynamics and mechanisms involved in the inflammation-induced by cigarette smoking. It must be recognized that localization of leukocytes at an inflammatory site in the lung involves three separate, but related, didactic phenomena: (1) recruitment (a multi-step process in itself including circulatory cytosis from bone marrow or spleen, adhesion and diapedesis to and through endothelial surfaces); (2) proliferation, persistence and activity state at the site (complicated by cell-cell interactions and feed-forward mechanisms); and (3) resolution (by diapedesis into the airway lumen, apoptosis, phagocytosis and rare return to the circulation). In man, studies investigating the effects of acute smoke exposure have been limited. Of the approximately 25 reports in the literature, 16 look for inflammatory changes mostly by sampling the blood within a couple hours of a single exposure period (reviewed in [60]). Its been reported that smoke increased numbers of neutrophils in the blood within an hour, increased the marginated pool of neutrophils in the lung microvasculature, and many, but not all studies, demonstrate an increase in markers of oxidative damage. In rodents, acute exposure to cigarette smoke (typically 3 consecutive days of exposure) increases inflammatory cytokine and chemokine levels in the lung tissue and BALF [57, 61, 62]. This is followed first by neutrophils and subsequently by macrophages and lymphocytes infiltrating the airspaces in most species and strains examined [63, 64]. Increases in elastin and collagen breakdown products in BALF after a single exposure to smoke have also been reported, suggesting that there are a number of potential mechanisms that may be driving the inflammation in this model [65]. In addition to these inflammatory changes, there is also an increase in oxidative damage to proteins and lipids [66–68] and, in rats, there is also an increase in the number of mucus containing epithelial cells in the central airways [62, 69]. A model with a slightly longer exposure regimen (2–4 weeks), some have termed sub-chronic, has also been used to study the response to smoke inhalation. The added advantage of this model is that there are greater numbers of macrophages and lymphocytes measurable in the BALF, and it allows compounds to be tested using therapeutic dosing regimens [69, 70].

These shorter duration models are ideal for profiling pharmacological tools and candidate compounds. Such studies have provided important insights into the mechanisms orchestrating the inflammation due to smoke exposure. Unlike other models of lung injury (e.g., lipopolysaccharide-induced inflammation), the inflammation in acute smoking models is not sensitive to glucocorticoid treatment [56, 64, 66]. Steroids fail to decrease the number of infiltrating neutrophils and macrophages in both rats [64, 66] and mice [56]. One possible explanation for this lack of effect is due to a decrease in histone deacetylase (HDAC) activity [71]. Histone acetylases (HATs) and HDACs work in concert to help turn on and off inflammatory gene transcription, respectively. After steroids bind the glucocorticoid receptor and it translocates to the DNA, this complex recruits HDACs to shut off gene transcription. It has been demonstrated that smoke exposure causes oxidant-mediated damage to HDACs (specifically, HDAC-2) in macrophages from COPD patients, and this is thought to cause the decreased activity of these enzymes. Similarly, rats exposed to smoke in these acute models also have decreased HDAC-2 activity [66].

Phosphodiesterase 4 (PDE4) is the principal phosphodiesterase isoform present in inflammatory cells. Inhibitors serve as anti-inflammatory agents by causing an increase in intracellular cAMP levels, resulting in the de-activation of immune cells present at the site of injury. Cilomilast and roflumilast are two PDE4 inhibitors that have demonstrated some potential clinical benefit by improving lung function [72, 73] and inflammation [74] in COPD patients. Roflumilast has also been shown to be effective at reducing neutrophil infiltration and oxidant damage after acute smoke exposure [56, 75]. Thus, PDE4 inhibitors together with steroids can act as positive and negative control compounds, respectively, to help validate these models.

Neutrophils increase in the airways after acute smoke exposure in a significant fashion and are major contributors to lung cell injury through their ability to release cytotoxic humoral mediators including oxygen-derived free radicals, proteases, cytokines and chemokines. Several neutrophil chemoattractants have been shown to increase after acute smoke exposure, indicating there may be multiple mechanisms driving this acute response. CXCR2 ligands in particular, have been shown to precede the infiltration of neutrophils in the BALF after exposure and therefore, are thought to be important chemoattractants [62, 76]. This hypothesis was further supported by data showing that CXCR2 antagonists, SB332235 and SCH-N, partially inhibit the neutrophil infiltration in response to smoke in both rats [62] and mice [57], respectively. CXCR2 ligands are also said to affect mucus production [77]. In rat, SB332235 did reduce smoke-induced goblet cell metaplasia at the lowest dose tested, but showed no efficacy at higher doses [62]. The reason for this anomaly remains unclear.

Oxidant stress is also believed to play a large role in disease pathogenesis and contribute to the enhanced inflammation observed in COPD patients [78]. This suggestion is further strengthened by data showing that a catalytic antioxidant mimetic, AEOL 10150, effectively reduces the inflammation and oxidant damage in the lung after 2 weeks of smoke exposure in rats [70]. Prophylactic and therapeutic administration of N-acetylcysteine also reduced smoke-induced mucus hypersecretion in the rat ([69]; and Danahay et al., unpublished observations). The inflammation induced by reactive oxygen species is believed to be due to the activation of the p38 MAP kinase pathway [46] and NF-κB [47, 79], leading to the production of pro-inflammatory cytokines and chemokines. SB239063, a p38 inhibitor, has also been reported to inhibit the inflammation in an acute smoking model in mice (Tralau-Stewart et al. reported in [80]), suggesting this pathway is involved in mediating the acute response to smoke inhalation.

The expression of TNF-α and IL-1β are increased after acute smoke exposure and these two broad spectrum inflammatory cytokines also activate the p38 pathway [61, 62, 66]. TNF-α and IL-1β contribute to the inflammatory process by inducing the expression of additional cytokines, chemokines, and proteases as well as activating the endothelium, leading to leukocyte migration, adhesion, and diapedesis [55]. Both of these cytokines have been shown to play key roles in mediating the acute effects of smoke inhalation. An antibody to IL-1β attenuated the acute inflammatory response to smoke [81] and mice deficient for both TNF-α receptors had reduced numbers of infiltrating immune cells and less evidence of matrix degradation in lavage fluid [61].

Proteases are known to produce chemotactic matrix fragments that are believed to drive at least part of the inflammatory response in smokers [27–29, 76]. Data from studies investigating the effects of protease inhibitors in these acute smoking models support these claims. Broad spectrum matrix metalloprotease (MMP) inhibitors have been shown to reduce both smoke-induced matrix degradation and lung inflammation ([65]; Fitzgerald et al. reported in [80]). Similarly, serine protease inhibitors as well as α1-antitrypsin protein also significantly inhibited the matrix destruction and acute inflammatory response to smoke inhalation [65, 82]. In addition to creating chemotactic matrix fragments, MMPs, and in particular macrophage metalloelastase (MMP-12), are also important activators of latent cytokines and chemokines. Therefore, this could be an additional mechanism through which proteases and in particular MMPs can contribute to the disease process.

This aspect of MMP activity was examined using genetically modified mice. In an elegant series of experiments, Churg and colleagues [65, 82] demonstrated that acute smoke exposure induced neutrophil infiltration in the lung, and proteases from these cells were responsible for the matrix degradation observed in this model. Subsequently, they illustrated that smoke-induced neutrophil infiltration was blocked in mice deficient for MMP-12 [55]. Their data showed the NF-κB pathway was activated after smoke exposure in both wild-type and MMP-12$^{-/-}$ mice, and some inflammatory mediators were up-regulated in both mouse lines. However, the major difference was that TNF-α levels were not increased in the MMP-12$^{-/-}$ mice after smoke exposure. They demonstrated *in vitro* using macrophages from the knockout mice and wild-type mice in the presence and absence of a broad spectrum MMP inhibitor that TNF-α release from macrophages was dependent on MMP-12 activity. These data, along with the data from the study describing the effects in mice deficient for both TNF-α receptors [61] allowed them to conclude that one of principle contributions of MMP-12 to the acute inflammatory response after short-term exposure to smoke was its ability activate latent TNF-α. Once activated, TNF-α mediated neutrophil infiltration in the lung, at least in part, through its activation of the endothelium. More specifically, TNF-α induced the expression of E-selectin on the endothelial wall, which allows circulating neutrophils to stick and subsequently infiltrate the lung airspaces [55].

Unfortunately, recent clinical studies investigating the effects of an anti-TNF-α therapy infliximab in COPD patients have shown no efficacy [83, 84]. This suggests that although there may be good correlation between the effects of certain compounds in these models and the clinic (i.e., steroids and PDE4 inhibitors), they may not always be predictive of clinical effect. However, further studies using this approach, possibly in end-stage patients, need to be conducted before definitive conclusions about the efficacy of anti-TNF-α therapies in COPD can be drawn.

Chronic smoking models

The effects of chronic smoke exposure have been investigated in many species with the majority of the work being done in guinea pig, mouse, and rat. The major features of these models include marked inflammatory lesions that resolve very slowly, increased mucus production, epithelial hyperplasia, and the development of emphysema (Fig. 3A–E). The emphysematous lesions in these models are due to increases in alveolar duct area and enlarged alveolar spaces. Fewer studies in rodents have looked at the effect of chronic smoke exposure on small airway structure and function; however, there are a few reports of smoke-induced increases in airway fibrosis and changes in airway function [85–88]. However, while most of the inflammatory and structural alterations in these chronic rodent smoking models are similar to those that occur in chronic smokers, they are subtle changes by comparison to those observed in COPD patients. As such, few studies have reported any significant change in the lung function of smoke-exposed animals. It is likely that this is because these models (typically run over 4–12 months) do not last long enough to mimic the extent of damage done over the 40+ years a typical COPD patient smokes. In addition, these smoking models take place under specific-pathogen free environments. This keeps the animals free of any respiratory infections that are likely to be a key element in the progression of COPD.

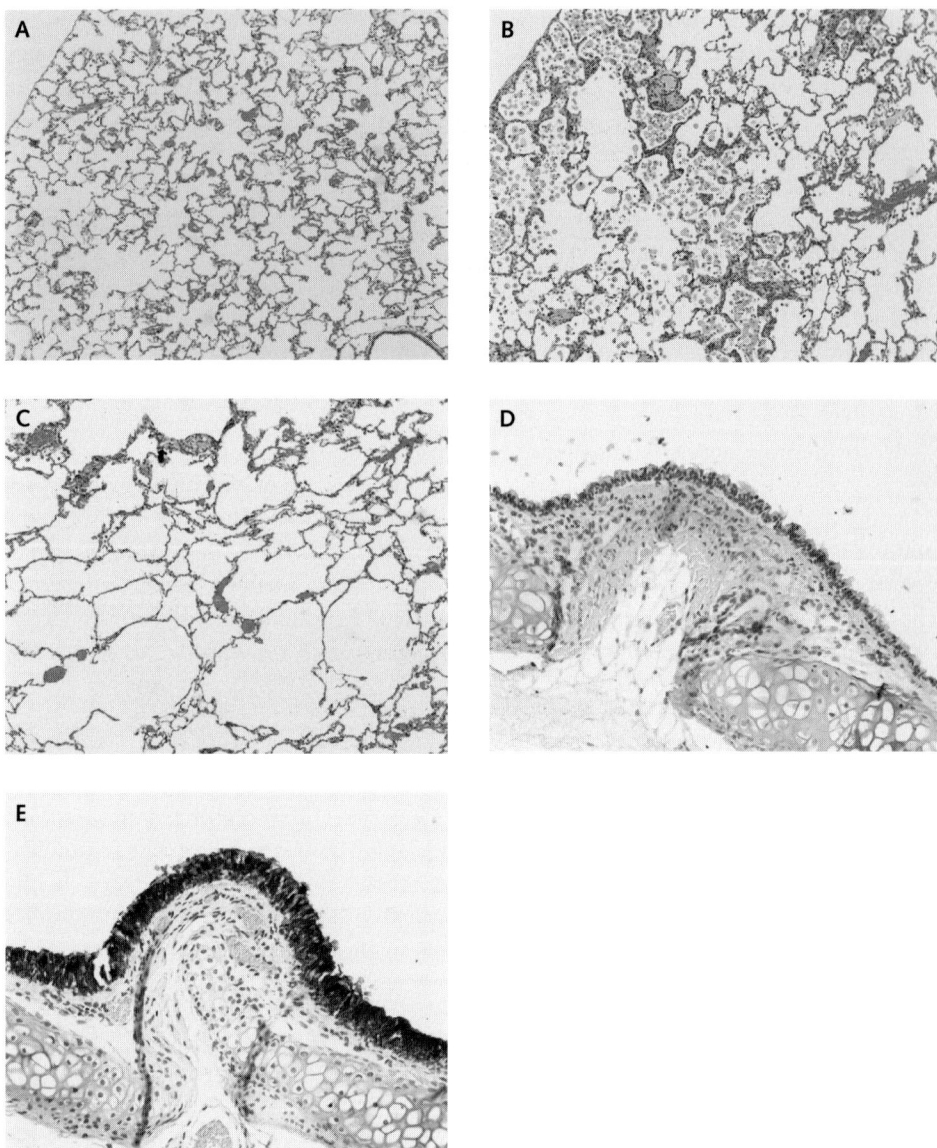

Figure 3
Images of lung sections from air-exposed (A and D) and smoke-exposed (B, C, and E) rats.
Smoke exposure leads to inflammatory cell aggregates in the alveolar spaces (B), and this is
believed to lead to the eventual loss of alveolar integrity (C). Another prominent feature of
this model is the thickening of the epithelium and goblet cell metaplasia that occurs after
smoke exposure (E).

The guinea pig appears to be the species most susceptible to the developing emphysema-like changes after chronic smoke exposure. The model described by Wright and colleagues [89] in the guinea pig is the first that resembled many of the pathological and physiological changes consistent with COPD. Cigarette smoke exposure caused a progressive increase in airspace enlargement and a decrease in lung function, which was observable after 3 months of exposure and did not resolve after smoking cessation. Associated with these changes were increased numbers of neutrophils, macrophages and lymphocytes (predominantly CD4+ T cells) [90, 91]. This inflammatory cell infiltrate was present in both the parenchyma and airway wall [90, 91]; however, there was limited remodeling of the airway observed [92, 93]. The only observable change to the smaller airways was increased secretory cell metaplasia in smoke-exposed guinea pigs compared to air-exposed controls [93].

Similar inflammatory and emphysema-like changes have been reported in mice after 4–6 months of smoke exposure. There is a progressive increase in neutrophils, macrophages, lymphocytes and dendritic cells, a loss of ciliated epithelial cells [63, 94], an increase in mucus containing epithelial cells [53], occasional obstruction of small airways by inflammatory cells and debris [94], and emphysematous destruction of the lung [52, 95]. One limitation of mouse models of lung disease in general is the differences in lung function and anatomy between mouse and man. Mice are obligate nose-breathers, have much less extensive branching of the airways, and lack submucosal glands (except in the trachea), lack true goblet cells, have no cough reflex, and respiratory function is more difficult to assess. Another limitation, mentioned previously, is the significant differences in the sensitivity of certain strains for developing these changes [53, 54, 59]. While this is a limitation to modeling in one respect, it is an advantage being exploited by some to try and identify genes that confer disease susceptibility and resistance [54, 96]. The mouse genome is well-defined and methods for gene manipulation in mice are well established, making them an ideal species to investigate the molecular aspects of chronic smoke-induced changes.

Rats are perhaps the most resistant species to smoke-induced airspace enlargement, but possibly the best for examining smoke-induced mucus production [51, 52]. Chronic smoke exposure induces progressive increases in neutrophils, macrophages, and lymphocytes, a marked increase in the number of goblet cells in both the central and smaller airways [64, 69], mild airway remodeling [64, 85, 86], and airspace enlargement, but this remains controversial. There are conflicting reports on whether emphysema-like changes can be generated in rats after chronic smoke exposure [52, 64, 97–99]. This may be due to exposure and strain differences or the robust regenerative capacity of the rat lung, which is not fully developed at birth. (The epiphyses never close [100] and, therefore, thoracic growth and alveolarization persist throughout life [101].) Although, several groups have demonstrated airspace enlargement, none have found a corresponding change in lung function [64, 99].

Consistent with observations from the acute models, there are increased levels of cytokines, chemokines, matrix degradation, and oxidative damage in lungs of chronically exposed animals, all of which have been implicated in contributing to the inflammation and resulting pathologies associated with these models. Most of the pharmacological and molecular investigations in these chronic models have been performed in mice because of the ease of housing, generating airspace enlargement, and genetic manipulation. The most common strain used is the c57BL/6, which develops emphysema between 4 and 6 months. It is believed that one of the reasons this strain is more susceptible to smoke-induced pathology is because they have lower circulating levels of α1-antitrypsin proteinase inhibitor [59].

Interestingly, the anti-inflammatory effects of many of the compounds and gene deletions tested in the acute models have been effective at attenuating the inflammation and pathologies that occur in the chronic smoking models. For instance, oral administration of roflumilast protected mice from smoke-induced inflammation, oxidant stress, and emphysema [75, 102], while budesonide did not [102]. Oxidative damage to HDAC was again the mechanism proposed for the lack of steroid efficacy [66]. Oxidant stress has also been demonstrated to play a key role in mediating the chronic effects of smoke-induced lung damage. As mentioned previously, in addition to activating inflammatory pathways (e.g., p38 MAP kinase and NF-κB activation), oxidants can also inactivate endogenous protease inhibitors, and thereby potentiate the proteolytic activity in the lung that contributes to airspace enlargement. Similar to the elastase model, Nrf2-deficient mice had greater lung inflammation and emphysematous destruction in response to 6 months of smoke exposures [103]. As in the sub-chronic model, the catalytic anti-oxidant AEOL 10150 attenuated inflammation after 2 months of smoke exposures [70] and N-acetylcysteine protected against smoke-induced small airway remodeling in rat [85]. Further, overexpression CuZn-superoxide dismutase in mice attenuated smoke-induced inflammation, oxidative damage, protease expression, and completely prevented emphysematous destruction of the lung [68].

Inflammatory cytokines also play an important role in the development of airspace enlargement. IFN-γ can induce the expression of CCR5 ligands MIP-1a, MIP-1b, and RANTES, as well as enzymes involved in lung matrix degradation (MMP-9 and MMP-12) and apoptosis. Ma and colleagues [104] demonstrated that all of these mediators were increased in the lungs of smoke-exposed mice that developed emphysema. They also showed that mice deficient in IFN-γ or CCR5 were protected from smoke-induced airspace enlargement [104]. Further, mice that overexpressed IFN-γ developed the same types of pathologies associated with COPD such as airway fibrosis, mucus hypersecretion, and emphysema [105], suggesting that IFN-γ is both necessary and sufficient to induce emphysema. As in the acute model, animals deficient for both TNF-α receptors were also protected from smoke-induced inflammation and lung damage. Initial studies showed that airspace enlargement was reduced by approximately 70% in the double-knockout mice compared to wild-type controls [58]. Subsequent studies using specific TNF-α R1 or R2 knockout mice suggest that TNF-α R2 may be the more important of the two receptors for mediating the lung destruction associated with chronic smoke exposure [106]. Again, mice deficient in TNF-α R2 had much less inflammation and protease expression compared to wild-type and TNF-α R1 knockout mice.

Proteases have been shown to be key to the development of smoke-induced emphysema. MMP inhibitors and serine protease inhibitors significantly ameliorate the airspace enlargement in both guinea pigs and mice [107–110]. In particular, MMP-12 appears to be centrally involved, as mice deficient for MMP-12 were completely protected from cigarette smoke-induced emphysema, and this has been repeated on multiple background strains [95, 111]. As in the acute model, MMP-12 contributes to these changes through its ability to degrade lung elastin, generate chemotactic elastin fragments, and activate latent cytokines, leading to greater inflammatory cells into the lung [29, 55]. MMP-12$^{-/-}$ mice had less macrophages in the lungs after smoke exposure compared to wild types, but there was no effect on the numbers of neutrophils [95]. The macrophage appeared to be the predominant inflammatory cell in the lung after smoking, as increases in lung neutrophil numbers were much lower. However, although ablation of the MMP-12 gene was sufficient for complete protection against the effects of smoke, it appears that both macrophage metalloelastase and NEs are important in generating airspace enlargement.

Shapiro and colleagues [111] demonstrated that mice deficient for the NE gene were also partially protected smoking-induced emphysema by approximately 60%. As in the acute smoking model [65, 82], blocking NE activity reduced the number of neutrophils in the BALF and macrophages in the lung after smoke exposure [111]. It was proposed that NE was important for neutrophil migration by cleaving the CD11/CD18-ICAM complex that keeps it tethered to the endothelium [112]. Further, NE is expressed on the surface of pro-inflammatory monocytes infiltrating tissue, thus deletion of this enzyme reduced the numbers of monocytes able to get into the lung [113]. However, the most important finding in this study was the illustration of the interdependence of proteases in regulating one another's activity. NE was shown to degrade the endogenous inhibitor of MMP-12, TIMP-1, and likewise, MMP-12 was shown to degrade α1-antitrypsin, thus enhancing each other's potency [111]. In addition, it was discovered that NE was required for the proteolytic activation of pro-MMP-12. The conclusion drawn from these experiments was that the contribution of NE to smoke-induced airspace enlargement was secondary to its effects on lung elastin [111]. Rather it was its role in directing monocyte transvascular migration into the lung and its activation of latent MMP-12. The authors' conveyed their surprise by these findings, as they did not expect neutrophils and NE to play a role in generating airspace enlargement in this model. They explained that in human disease, where there is greater incidence of viral and bacterial infection, neutrophils are likely to be more involved; however, in mice that are exposed to smoke in a specific pathogen free environment, neutrophils were not expected to be as important [111]. Further, it had been previously reported that depleting neu-trophils in rats had no effect on smoke-induced airspace enlargement, while deple-tion of macrophages did provide protection [114].

All of these findings suggest that there is a close relationship between oxidants, inflammation, and proteolytic activity which contributes to the lung destruction generated from cigarette smoking. Again, the major limitation is that the inflam-mation and pathologies are subtle and, at best, may only reflect the early stages of the disease. However, there is good correlation between the acute, sub-chronic and chronic smoking models as well as some valuable parallels to the clinic. This allows investigators to first examine mechanisms and conduct compound profiling in the shorter duration models to obtain valuable data and some confidence in their approach before attempting such studies in the chronic, more labor-intensive mod-eling systems.

Future directions

Modeling exacerbations and co-morbidities

As the need for therapeutic approaches for COPD is immediate, many patients require agents to improve their quality of life in a disease which has already mani-fested itself as a presumably irreversible state. In mid-to-late stage disease, acute exacerbations by stress, exercise or viral and bacterial infections commonly occur and contribute to disease progression. At this stage, the adaptive immune response appears to play a larger role and is more closely associated with lung function decline and survival prognosis than the cells involved in the innate immune response [8]. The models described above appear to only require the innate immune response to drive these changes [115]. Few have tried to model these later stage changes by combining both cigarette smoke with viral or bacterial challenges [90, 116, 117]. Meshi and colleagues [90] combined cigarette smoke exposure with a latent aden-oviral infection in guinea pigs. They demonstrated an increase in cells involved in adaptive immunity and associated with these changes was greater airspace enlarge-ment. Therefore, it may be necessary to combine these two stimuli to model late-stage disease. In addition, as COPD progresses, patients are afflicted with a number of co-morbidities that affects their quality of life. Skeletal muscle wasting, in partic-ular, is associated greater disability and a powerful predictor of mortality [118]. Sev-eral groups are also now investigating this aspect of the disease using chronic smok-ing models [106, 119]. In addition, exercise testing in rodents may provide a method which could act as a potential surrogate quality of life measure similar to the 6-min walk test in man [120]. Remodeling of the pulmonary vasculature in COPD patients leads to vascular dysfunction and secondary pulmonary hypertension [121]. These changes, which also affect quality of life and are associated with increased mortality, have been modeled in guinea pigs using cigarette smoke [122]. Such models can be used to help identify therapies targeted at halting or reversing the systemic aspects of this disease. Treating the co-morbidities could possibly have a more substantial impact on improving the lives and life expectancy of advanced COPD patients than therapies directed at the disease in the airways.

Transgenic models of COPD

Animal research in chronic obstructive pulmonary disease has proceeded in two general directions, which are not mutually exclusive: (1) models to help discover and describe the etiology of the disease, and (2) models in which to test therapeutic agents. Both approaches lend a hand to each other by helping to discern new targets for therapeutic research and by providing pharmacological tools to test hypotheses. Methods that allow for gene deletion or overexpression in mice have advanced significantly over the last decade and these systems are extremely useful for testing hypothesis about the role of certain genes in specific diseases as illustrated above. There have been three recently described transgenic mouse models that resemble many of the pathological aspects of COPD, and these mice are also now being used to investigate mechanisms potentially responsible for disease and profile candidate compounds. Inducible, targeted overexpression of IFN-γ, IL-13, and human IL-1β in adult mouse lung [using the Clara cell 10-kDa-reverse tetracycline transactivator (CC10-rtTA) promoter construct] resulted in marked inflammatory lesions in the lung, mucus cell metaplasia, significant airway fibrosis, and emphysema [105, 123, 124]. Associated with these changes was a significant increase in the expression of many proteases, including MMP-9 and MMP-12. Again, the cycle of inflammatory and resident cell production of these proteases and proteolytic activity leading to greater inflammation is represented in these models. MMP-9 and MMP-12, in particular, were implicated in driving the developing pathologies. To test this hypothesis, Lanone and colleagues [125] overexpressed IL-13 in the lungs of mice deficient for MMP-9 and MMP-12. Airspace enlargement did not occur in IL-13$^{+/+}$-MMP-9$^{-/-}$ or IL-13$^{+/+}$-MMP-12$^{-/-}$ mice. In addition, IL-13$^{+/+}$ mice develop progressive fibrodestructive lung alterations that result in a pre-mature death. Survival was enhanced in IL-13$^{+/+}$-MMP-9$^{-/-}$ and IL-13$^{+/+}$-MMP-12$^{-/-}$ mice. Thus, these systems provide further opportunities for testing novel hypothesis and examining how specific mediators may contribute to disease pathogenesis.

Conclusion

Concerning pharmacological characterization of drugs which may be effective in the treatment of COPD, just as there is no single animal model, we must also remember that no single drug has provided consistent efficacy in the clinical treatment of the disease. Therefore, a potential therapeutic drug regimen must be assessed in a model which appropriately reflects a particular aspect of the disease (i.e., inflammatory cell infiltration, mucus hypersecretion, airway wall remodeling, emphysema, and systemic disease). Because some agents (agonists, inhibitors or antagonists) may work in only certain animals, the appropriate stimulus and the particular therapeutic drug standard to which it should be compared may be species and strain dependent. Therefore, whenever possible, careful *in vitro* or *ex vivo* coordination and comparison of the activity found in tissues from the species and models selected to healthy and diseased human tissues should be made. Although rational therapeutic approaches based on inhibitory activity in a number of these models may increase the level of confidence in finding efficacy in the disease state, one should not oversimplify the etiology of the disease to fit the overall profile of the drug.

Acknowledgements

The authors would like to especially thank Dr. Mark Birrell for his help in the preparation of this manuscript and helpful insights. The author would also like to thank Dr. Sissie Wong, Dr. Elizabeth Hardaker, Dr. Michael Salmon, and Prof. Maria Belvisi.

References

1   Pauwels, RA Buist, S, Calverley, PMA, Jenkins, CR, Hurd, SS (2001) Global strategy for the diagnosis, management, and prevention of chronic obstructive pulmonary disease. *Am J Respir Crit Care Med* 163: 1256–1276

2   Higgins BG, Francis HC, Yates CJ, Warburton CJ, Fletcher AM, Reid JA, Pickering CA, Woodcock AA (1995) Effects of air pollution on symptoms and peak expiratory flow measurements in subjects with obstructive airways disease. *Thorax* 50: 149–155

3   Jany B, Gallup M, Tsuda T, Basbaum C (1991) Mucin gene expression in rat airways following infection and irritation. *Biochem Biophys Res Commun* 181: 1–8

4   Lei YH, Barnes PJ, Rogers DF (1995) Mechanisms and modulation of airway plasma exudation after direct inhalation of cigarette smoke. *Am J Respir Crit Care Med* 151: 1752–1762

5   Dusser DJ, Djokic TD, Borson DB, Nadel JA (1989) Cigarette smoke induces bronchoconstrictor hyperresponsiveness to substance P and inactivates airway neutral endopeptidase in the guinea pig. Possible role of free radicals. *J Clin Invest* 84: 900–906

6   Turato G, Zuin R, Miniati M, Baraldo S, Rea F, Beghe B, Monti S, Formichi B, Boschetto P, Harari S et al (2002) Airway inflammation in severe chronic obstructive pulmonary disease: relationship with lung function and radiologic emphysema. *Am J Respir Crit Care Med* 166: 105–110

7   Saetta M, Di Stefano A, Turato G, Facchini FM, Corbino L, Mapp CE, Maestrelli P, Ciaccia A, Fabbri LM (1998) CD8+ T-lymphocytes in peripheral airways of smokers with chronic obstructive pulmonary disease. *Am J Respir Crit Care Med* 157: 822–826

8   Hogg JC, Chu F, Utokaparch S, Woods R, Elliott WM, Buzatu L, Cherniack RM, Rogers RM, Sciurba FC, Coxson HO, Pare PD (2004) The nature of small-airway obstruction in chronic obstructive pulmonary disease. *N Engl J Med* 350: 2645–2653

9   Hogg JC (2004) Pathophysiology of airflow limitation in chronic obstructive pulmonary disease. *Lancet* 364: 709–721

10 van der Strate BW, Postma DS, Brandsma CA, Melgert BN, Luinge MA, Geerlings M, Hylkema MN, van den Berg A, Timens W, Kerstjens HA (2006) Cigarette smoke-induced emphysema: A role for the B cell? *Am J Respir Crit Care Med* 173: 751–758

11 Voelkel N, Taraseviciene-Stewart L (2005) Emphysema: an autoimmune vascular disease? *Proc Am Thorac Soc* 2: 23–25

12 Taraseviciene-Stewart L, Scerbavicius R, Choe KH, Moore M, Sullivan A, Nicolls MR, Fontenot AP, Tuder RM, Voelkel NF (2005) An animal model of autoimmune emphysema. *Am J Respir Crit Care Med* 171: 734–742

13 American Lung Association (2005) COPD Fact Sheet. www.lungusa.com, Home > Chronic Obstructive Pulmonary Disease (COPD) center > COPD Fact Sheet (http://www.lungusa.org/site/pp.asp?c=dvLUK9O0E&b=35020).

14 Fletcher C, Peto R (1977) The natural history of chronic airflow obstruction. *Br Med J* 25: 1645–1648

15 Lundback B, Lindberg A, Lindstrom M, Ronmark E, Jonsson AC, Jonsson E, Larsson LG, Andersson S, Sandstrom T, Larsson K (2003) Obstructive lung disease in Northern Sweden studies. Not 15 but 50% of smokers develop COPD? – Report from the Obstructive Lung Disease in Northern Sweden Studies. *Respir Med* 97: 115–122.

16 Eriksson S (1965) Studies in α1-antitrypsin deficiency. *Acta Med Scand* (Suppl) 432: 1–85

17 Laurell C-B, Eriksson S (1963) The electrophoretic α1-globulin pattern of serum in α1-antitrypsin deficiency. *Scand J Clin Lab Invest* 15: 132–140

18 Gross P, Pfitzer EA, Tolker E, Babyak MA, Kaschak M (1965) Experimental emphysema. Its production with papain in normal and silicotic rats. *Arch Environ Health* 11: 50–58

19 Thurlbeck WM (1990) Pathology of chronic obstructive pulmonary disease. *Clin Chest Med* 11: 389–403

20 Snider GL (1992) Emphysema: The first two centuries and beyond. A historical overview, with suggestions for future research: part 2. *Am Rev Respir Dis* 146: 1615–1622

21 Snider GL, Lucey EC, Stone PJ (1986) Animal models of emphysema. *Am Rev Respir Dis* 133: 149–169

22 Husain AN, Kumar V (2005) The lung. In: Kumar V, Abbas AK, Fausto N (eds): *Robbins and Cotran – Pathologic basis of disease*. Elsevier Saunders, Philadelphia, 711–772

23   Bates DV, Macklem PT, Christie RV (1971) *Respiratory function in disease*. WB Saunders, Toronto

24   Lucey EC, Stone PJ, Christensen TG, Breuer R, Snider GL (1988) An 18-month study of the effects on hamster lungs of intratracheally administered human neutrophil elastase. *Exp Lung Res* 14: 671–686

25   Birrell MA, Wong S, Hele DJ, McCluskie K, Hardaker E, Belvisi MG (2005) Steroid-resistant inflammation in a rat model of chronic obstructive pulmonary disease is associated with a lack of nuclear factor-kappaB pathway activation. *Am J Respir Crit Care Med* 172: 74–84

26   Chang C, Houck JC (1970) Demonstration of the chemotactic properties of collagen. *Proc Soc Exp Biol Med* 134: 22–26

27   Postlethwaite AE, Kang AH (1976) Collagen and collagen peptide-induced chemotaxis of human blood monocytes. *J Exp Med* 143: 1299–1307

28   Senior RM, Griffin GL, Mecham RP (1980) Chemotactic activity of elastin-derived peptides. *J Clin Invest* 66: 859–862

29   Houghton AM, Quintero PA, Perkins DL, Kobayashi DK, Kelley DG, Marconcini LA, Mecham RP, Senior RM, Shapiro SD (2006) Elastin fragments drive disease progression in a murine model of emphysema. *J Clin Invest* 116: 753–759

30   Kaplan PD, Kuhn C, Pierce JA (1973) The induction of emphysema with elastase. I. The evolution of the lesion and the influence of serum. *J Lab Clin Med* 82: 349–356

31   Kuhn C, Yu SY, Chraplyvy M, Linder HE, Senior RM (1976) The induction of emphysema with elastase. II. Changes in connective tissue. *Lab Invest* 34: 372–380

32   Massaro GD, Massaro D (1997) Retinoic acid treatment abrogates elastase-induced pulmonary emphysema in rats. *Nat Med* 3: 675–677

33   Massaro GD, Massaro D (1996) Postnatal treatment with retinoic acid increases the number of pulmonary alveoli in rats. *Am J Physiol* 270: L305–310

34   Massaro D, Massaro GD (1986) Dexamethasone accelerates alveolar wall thinning and alters wall composition. *Am J Physiol* 251: R218–R224

35   Ong DE, Chytil F (1976) Changes in levels of cellular retinol- and retinoic-acid-binding protein of liver and lung during perinatal development of rat. *Proc Natl Acad Sci USA* 73: 3976–3978

36   Grummer MA, Zachman RD (1995) Postnatal rat lung retinoic acid receptor (RAR) mRNA expression and effects of dexamethasone on RAR-bmRNA. *Pediatr Pulmonol* 20: 234–240

37   Massaro D, Teich N, Maxwell S, Massaro GD, Whitney P (1985) Postnatal development of alveoli: Regulation and evidence for a critical period in rats. *J Clin Invest* 76: 1297_1305

38   Fujita M, Ye Q, Ouchi H, Nakashima N, Hamada N, Hagimoto N, Kuwano K, Mason RJ, Nakanishi Y (2004) Retinoic acid fails to reverse emphysema in adult mouse models. *Thorax* 59: 224–230

39   Lucey EC, Goldstein RH, Breuer R, Rexer BN, Ong DE, Snider GL (2003) Retinoic acid

does not affect alveolar septation in adult FVB mice with elastase-induced emphysema. *Respiration* 70: 200–205

40   March TH, Cossey PY, Esparza DC, Dix KJ, McDonald JD, Bowen LE (2004) Inhalation administration of all-trans-retinoic acid for treatment of elastase-induced pulmonary emphysema in Fischer 344 rats. *Exp Lung Res* 30: 383–404

41   Birrell MA, Wong S, Hardaker E, Catley MC, McCluskie K, Collins M, Haj-Yahia S, Belvisi MG (2006) I{kappa}B kinase-2 independent and dependent inflammation in airway disease models: relevance of IKK- 2 inhibition to the clinic. *Mol Pharmacol* [Epub ahead of print]

42   Lucey EC, Keane J, Kuang PP, Snider GL, Goldstein RH (2002) Severity of elastase-induced emphysema is decreased in tumor necrosis factor-alpha and interleukin-1beta receptor-deficient mice. *Lab Invest* 82: 79–85

43   Rubio ML, Martin-Mosquero MC, Ortega M, Peces-Barba G, Gonzalez-Mangado N (2004) Oral N-acetylcysteine attenuates elastase-induced pulmonary emphysema in rats. *Chest* 125: 1500–1506

44   Shinohara T, Kaneko T, Nagashima Y, Ueda A, Tagawa A, Ishigatsubo Y (2005) Adenovirus-mediated transfer and overexpression of heme oxygenase 1 cDNA in lungs attenuates elastase-induced pulmonary emphysema in mice. *Hum Gene Ther* 16: 318–327

45   Ishii Y, Itoh K, Morishima Y, Kimura T, Kiwamoto T, Iizuka T, Hegab AE, Hosoya T, Nomura A, Sakamoto T et al (2005) Transcription factor Nrf2 plays a pivotal role in protection against elastase-induced pulmonary inflammation and emphysema. *J Immunol* 175: 6968–6975

46   Ogura M and Kitamura M (1998) Oxidant stress incites spreading of macrophages via extracellular signal-regulated kinases and p38 mitogen-activated protein kinase. *J Immunol* 161: 3569–3574

47   Rahman I, Gilmour PS, Jimenez LA, MacNee W (2002) Oxidative stress and TNF-alpha induce histone acetylation and NF-kappaB/AP-1 activation in alveolar epithelial cells: potential mechanism in gene transcription in lung inflammation. *Mol Cell Biochem* 234–235: 239–248

48   Carp H, Miller F, Hoidal JR, Janoff A (1982) Potential mechanism of emphysema: alpha-1-proteinase inhibitor recovered from lungs of cigarette smokers contains oxidized methionine and has decreased elastase inhibitor capacity. *Proc Natl Acad Sci USA* 79: 2041–2045

49   Pryor WA, Dooley MD, Church DF (1986) The mechanisms of inactivation of human alpha-1-proteinase inhibitor by gas phase cigarette smoke. *Free Radic Biol Med* 2: 161–168

50   Cantor JO, Cerreta JM, Keller S, Turino GM (1995) Modulation of airspace enlargement in elastase-induced emphysema by intratracheal instillment of hyaluronidase and hyaluronic acid. *Exp Lung Res* 21: 423–436

51   Wright JL, Churg A (2002) Animal models of cigarette smoke-induced COPD. *Chest* 122: 301S–306S

52   March TH, Barr EB, Finch GL, Hahn FF, Hobbs CH, Menache MG, Nikula KJ (1999)

Cigarette smoke exposure produces more evidence of emphysema in B6C3F1 mice than in F344 rats. *Toxicol Sci* 51: 289–299

53　Bartalesi B, Cavarra E, Fineschi S, Lucattelli M, Lunghi B, Martorana PA, Lungarella G (2005) Different lung responses to cigarette smoke in two strains of mice sensitive to oxidants. *Eur Respir J* 25: 15–22

54　Guerassimov A, Hoshino Y, Takubo Y, Turcotte A, Yamamoto M, Ghezzo H, Triantafillopoulos A, Whittaker K, Hoidal JR, Cosio MG (2004) The development of emphysema in cigarette smoke-exposed mice is strain dependent. *Am J Respir Crit Care Med* 170: 974–980

55　Churg A, Wang RD, Tai H, Wang X, Xie C, Dai J, Shapiro SD, Wright JL (2003) Macrophage metalloelastase mediates acute cigarette smoke-induced inflammation via tumor necrosis factor-alpha release. *Am J Respir Crit Care Med* 167: 1083–1089

56　Leclerc O, Lagente V, Planquois JM, Berthelier C, Artola M, Eichholtz T, Bertrand CP, Schmidlin F (2006) Involvement of MMP-12 and phosphodiesterase type 4 in cigarette Smoke-induced inflammation in mice. *Eur Respir J* 27: 1102–1109

57　Thatcher TH, McHugh NA, Egan RW, Chapman RW, Hey JA, Turner CK, Redonnet MR, Seweryniak KE, Sime PJ, Phipps RP (2005) Role of CXCR2 in cigarette smoke-induced lung inflammation. *Am J Physiol Lung Cell Mol Physiol* 289: L322–328

58　Churg A, Wang RD, Tai H, Wang X, Xie C, Wright JL (2004) Tumor necrosis factor-alpha drives 70% of cigarette smoke-induced emphysema in the mouse. *Am J Respir Crit Care Med* 170: 492–498

59　Cavarra E, Bartalesi B, Lucattelli M, Fineschi S, Lunghi B, Gambelli F, Ortiz LA, Martorana PA, Lungarella G (2001) Effects of cigarette smoke in mice with different levels of α1-proteinase inhibitor and sensitivity to oxidants. *Am J Respir Crit Care Med* 164: 886–890

60　van der Vaart H, Postma DS, Timens W, Hylkema MN, Willemse BW, Boezen HM, Vonk JM, de Reus DM, Kauffman HF, ten Hacken NH (2005) Acute effects of cigarette smoking on inflammation in healthy intermittent smokers. *Respir Res* 6: 22

61　Churg A, Dai J, Tai H, Xie C, Wright JL (2002) Tumor necrosis factor-alpha is central to acute cigarette smoke-induced inflammation and connective tissue breakdown. *Am J Respir Crit Care Med* 166: 849–854

62　Stevenson CS, Coote K, Webster R, Johnston H, Atherton HC, Nicholls A, Giddings J, Sugar R, Jackson A, Press NJ et al (2005) Characterization of cigarette smoke-induced inflammatory and mucus hypersecretory changes in rat lung and the role of CXCR2 ligands in mediating this effect. *Am J Physiol Lung Cell Mol Physiol* 288: L514–522

63　D'hulst AI, Vermaelen KY, Brusselle GG, Joos GF, Pauwels RA (2005) Time course of cigarette smoke-induced pulmonary inflammation in mice. *Eur Respir J* 26: 204–213

64　Stevenson CS, Winny C, Coote K, Giddings J, Whittaker P, Pohlmeyer-Esch G, Charman C, Danahay H, Butler K (2004) A chronic rat model of smoke-induced lung injury and comparison with an acute 24 h screening model. *Am J Respir Crit Care Med* 169: A205

65　Churg A, Zay K, Shay S, Xie C, Shapiro SD, Hendricks R, Wright JL (2002) Acute cigarette smoke-induced connective tissue breakdown requires both neutrophils and macrophage metalloelastase in mice. *Am J Respir Cell Mol Biol* 27: 368–374

66 Marwick JA, Kirkham PA, Stevenson CS, Danahay H, Giddings J, Butler K, Donaldson K, Macnee W, Rahman I (2004) Cigarette smoke alters chromatin remodeling and induces proinflammatory genes in rat lungs. *Am J Respir Cell Mol Biol* 31: 633–642

67 Churg A, Cherukupalli K (1993) Cigarette smoke causes rapid lipid peroxidation of rat tracheal epithelium. *Int J Exp Pathol* 74: 127–132

68 Foronjy RF, Mirochnitchenko O, Propokenko O, Lemaitre V, Jia Y, Inouye M, Okada Y, D'Armiento JM (2006) Superoxide dismutase expression attenuates cigarette smoke- or elastase-generated emphysema in mice. *Am J Respir Crit Care Med* 173: 623–631

69 Rogers DF and Jeffery PK (1986) Inhibition by oral N-acetylcysteine of cigarette smoke-induced "bronchitis" in the rat. *Exp Lung Res* 10: 267–283

70 Smith KR, Uyeminami DL, Kodavanti UP, Crapo JD, Chang LY, Pinkerton KE (2002) Inhibition of tobacco smoke-induced lung inflammation by a catalytic antioxidant. *Free Radic Biol Med* 33: 1106–1114

71 Ito K, Ito M, Elliott WM, Cosio B, Caramori G, Kon OM, Barczyk A, Hayashi S, Adcock IM, Hogg JC, Barnes PJ (2005) Decreased histone deacetylase activity in chronic obstructive pulmonary disease. *N Engl J Med* 352: 1967–1976

72 Compton CH, Gubb J, Nieman R, Edelson J, Amit O, Bakst A, Ayres JG, Creemers JP, Schultze-Werninghaus G, Brambilla C et al (2001) Cilomilast, a selective phosphodi-esterase-4 inhibitor for treatment of patients with chronic obstructive pulmonary disease: a randomised, dose-ranging study. *Lancet* 358: 265–270

73 Rabe KF, Bateman ED, O'Donnell D, Witte S, Bredenbroker D, Bethke TD (2005) Roflumilast – an oral anti-inflammatory treatment for chronic obstructive pulmonary disease: a randomised controlled trial. *Lancet* 366: 563–571

74 Gamble E, Grootendorst DC, Brightling CE, Troy S, Qiu Y, Zhu J, Parker D, Matin D, Majumdar S, Vignola AM et al (2003) Antiinflammatory effects of the phosphodi-esterase-4 inhibitor cilomilast (Ariflo) in chronic obstructive pulmonary disease. *Am J Respir Crit Care Med* 168: 976–982

75 Martorana PA, Beume R, Lucattelli M, Wollin L, Lungarella G (2005) Roflumilast fully prevents emphysema in mice chronically exposed to cigarette smoke. *Am J Respir Crit Care Med* 172: 848–853

76 Weathington NM, van Houwelingen AH, Noerager BD, Jackson PL, Kraneveld AD, Galin FS, Folkerts G, Nijkamp FP, Blalock JE (2006) A novel peptide CXCR ligand derived from extracellular matrix degradation during airway inflammation. *Nat Med* 12: 317–323

77 Miller AL, Strieter RM, Gruber AD, Ho SB, Lukacs NW (2003) CXCR2 regulates respiratory syncytial virus-induced airway hyperreactivity and mucus overproduction. *J Immunol* 170: 3348–3356

78 Repine JE, Bast A, Lankhorst I (1997) Oxidative stress in chronic obstructive pulmonary disease. Oxidative Stress Study Group. *Am J Respir Crit Care Med* 156: 341–357

79 Schreck R, Albermann K, Baeuerle PA (1992) Nuclear factor kappa B: an oxidative

stress-responsive transcription factor of eukaryotic cells (a review). *Free Radic Res Commun* 17: 221–237

80   Hele D (2002) First Siena International Conference on animal models of chronic obstructive pulmonary disease, Certosa di Pontignano, University of Siena, Italy, September 30–October 2, 2001. *Respir Res* 3: 12

81   Castro P, Legora-Machado A, Cardilo-Reis L, Valenca S, Porto LC, Walker C, Zuany-Amorim C, Koatz VL (2004) Inhibition of interleukin-1beta reduces mouse lung inflammation induced by exposure to cigarette smoke. *Eur J Pharmacol* 498: 279–286

82   Dhami R, Gilks B, Xie C, Zay K, Wright JL, Churg A (2000) Acute cigarette smoke-induced connective tissue breakdown is mediated by neutrophils and prevented by alpha1-antitrypsin. *Am J Respir Cell Mol Biol* 22: 244–252

83   van der Vaart H, Koeter GH, Postma DS, Kauffman HF, ten Hacken NH (2005) First study of infliximab treatment in patients with chronic obstructive pulmonary disease. *Am J Respir Crit Care Med* 172: 465–469

84   Antoniu SA (2006) Infliximab for chronic obstructive pulmonary disease: towards a more specific inflammation targeting? *Expert Opin Investig Drugs* 15: 181–184

85   Rubio ML, Sanchez-Cifuentes MV, Ortega M, Peces-Barba G, Escolar JD, Verbanck S, Paiva M, Gonzalez Mangado N (2000) N-Acetylcysteine prevents cigarette smoke induced small airways alterations in rats. *Eur Respir J* 15: 505–511

86   Cooper PR, Stevenson CS, Poll CT, Barnes PJ, Sturton RG (2005) Videomicroscopy of small airway function in precision cut lung slices prepared from tobacco smoke exposed rats. *Proc Amer Thor Soc* 2: A652

87   Wright JL, Sun JP, Churg A (1999) Cigarette smoke exposure causes constriction of rat lung. *Eur Respir J* 14: 1095–1099

88   Wang RD, Tai H, Xie C, Wang X, Wright JL, Churg A (2003) Cigarette smoke produces airway wall remodeling in rat tracheal explants. *Am J Respir Crit Care Med* 168: 1232–1236

89   Wright JL, Churg A (1990) Cigarette smoke causes physiologic and morphologic changes of emphysema in the guinea pig. *Am Rev Respir Dis* 142: 1422–1428

90   Meshi B, Vitalis TZ, Ionescu D, Elliott WM, Liu C, Wang XD, Hayashi S, Hogg JC (2002) Emphysematous lung destruction by cigarette smoke. The effects of latent adenoviral infection on the lung inflammatory response. *Am J Respir Cell Mol Biol* 26: 52–57

91   Selman M, Montano M, Ramos C, Vanda B, Becerril C, Delgado J, Sansores R, Barrios R, Pardo A (1996) Tobacco smoke-induced lung emphysema in guinea pigs is associated with increased interstitial collagenase. *Am J Physiol* 271: L734–743

92   James AL, Pare PD, Hogg JC (1988) Effects of lung volume, bronchoconstriction, and cigarette smoke on morphometric airway dimensions. *J Appl Physiol* 64: 913–919

93   Wright JL, Ngai T, Churg A (1992) Effect of long-term exposure to cigarette smoke on the small airways of the guinea pig. *Exp Lung Res* 18: 105–114

94   Shapiro SD (2000) Animal models for chronic obstructive pulmonary disease: age of klotho and marlboro mice. *Am J Respir Cell Mol Biol* 22: 4–7

95 Hautamaki RD, Kobayashi DK, Senior RM, Shapiro SD (1997) Requirement for macrophage elastase for cigarette smoke-induced emphysema in mice. *Science* 277: 2002–2004

96 Shapiro SD, Demeo DL, Silverman EK (2004) Smoke and mirrors: Mouse models as a reflection of human chronic obstructive pulmonary disease. *Am J Respir Crit Care Med* 170: 929–931

97 Wright JL, Sun JP, Vedal S (1997) A longitudinal analysis of pulmonary function in rats during a 12-month cigarette smoke exposure. *Eur Respir J* 10: 1115–1119

98 Heckman CA, Dalbey WE (1982) Pathogenesis of lesions induced in rat lung by chronic tobacco smoke inhalation. *J Natl Cancer Inst* 69: 117–129

99 Ofulue AF, Ko M, Abboud RT (1998) Time course of neutrophil and macrophage elastinolytic activities in cigarette smoke-induced emphysema. *Am J Physiol* 275: L1134–1144

100 Dawson AB (1934) Additional evidence of the failure of epiphyseal union in the skeleton of the rat: studies on wild and captive Norway rats. *Anat Rec* 60: 501–511

101 Thurlbeck WM (1975) Lung growth and alveolar multiplication. *Pathobiol Annu* 5: 1–34

102 Fitzgerald MF, Spicer D, Henning R (2003) Efficacy of the PDE4 inhibitor, BAY 19-8004 and a steroid in tobacco smoke models of pulmonary inflammation. *Am J Resp Crit Care Med* 167: A91

103 Rangasamy T, Cho CY, Thimmulappa RK, Zhen L, Srisuma SS, Kensler TW, Yamamoto M, Petrache I, Tuder RM, Biswal S (2004) Genetic ablation of Nrf2 enhances susceptibility to cigarette smoke-induced emphysema in mice. *J Clin Invest* 114: 1248–1259

104 Ma B, Kang MJ, Lee CG, Chapoval S, Liu W, Chen Q, Coyle AJ, Lora JM, Picarella D, Homer RJ, Elias JA (2005) Role of CCR5 in IFN-gamma-induced and cigarette smoke-induced emphysema. *J Clin Invest* 115: 3460–3472

105 Wang Z, Zheng T, Zhu Z, Homer RJ, Riese RJ, Chapman HA Jr, Shapiro SD, Elias JA (2000) Interferon gamma induction of pulmonary emphysema in the adult murine lung. *J Exp Med* 192: 1587–1600

106 D'hulst AI, Bracke KR, Maes T, De Bleecker JL, Pauwels RA, Joos GF, Brusselle GG (2006) Role of TNF{alpha} receptor 2 in cigarette smoke-induced pulmonary inflammation and emphysema. *Eur Respir J* 28: 102–112

107 Selman M, Cisneros-Lira J, Gaxiola M, Ramirez R, Kudlacz EM, Mitchell PG, Pardo A (2003) Matrix metalloproteinases inhibition attenuates tobacco smoke-induced emphysema in Guinea pigs. *Chest* 123: 1633–1641

108 Churg A, Wang RD, Xie C, Wright JL (2003) Alpha-1-Antitrypsin ameliorates cigarette smoke-induced emphysema in the mouse. *Am J Respir Crit Care Med* 168: 199–207

109 Wright JL, Farmer SG, Churg A (2002) Synthetic serine elastase inhibitor reduces cigarette smoke-induced emphysema in guinea pigs. *Am J Respir Crit Care Med* 166: 954–960

110 Fitzgerald MF, Spicer D, Fox C, Kobayashi DK, Shapiro S (2002) Effect of acute and

chronic smoke exposures to cigarette smoke in two strains of mice. Efficacy of a matrix metalloproteinase (MMP) inhibitor , BAY-15-7496, on emphysema development. *Am J Respir Crit Care Med* 165: A824

111 Shapiro SD, Goldstein NM, Houghton AM, Kobayashi DK, Kelley D, Belaaouaj A (2003) Neutrophil elastase contributes to cigarette smoke-induced emphysema in mice. *Am J Pathol* 163: 2329–2335

112 Loike JD, Sodeik B, Cao L, Leucona S, Weitz JI, Detmers PA, Wright SD, Silverstein SC (1991) CD11c/CD18 on neutrophils recognizes a domain at the N terminus of the A alpha chain of fibrinogen. *Proc Natl Acad Sci USA* 88: 1044–1048

113 Owen CA, Campbell MA, Boukedes SS, Stockley RA, Campbell EJ (1994) A discrete subpopulation of human monocytes expresses a neutrophil-like proinflammatory (P) phenotype. *Am J Physiol* 267: L775–785

114 Ofulue AF, Ko M (1999) Effects of depletion of neutrophils or macrophages on development of cigarette smoke-induced emphysema. *Am J Physiol* 277: L97–105

115 D'hulst AI, Maes T, Bracke KR, Demedts IK, Tournoy KG, Joos GF, Brusselle GG (2005) Cigarette smoke-induced pulmonary emphysema in scid-mice. Is the acquired immune system required? *Respir Res* 6: 147

116 Gualano RC, Vlahos R, Anderson GP (2006) What is the contribution of respiratory viruses and lung proteases to airway remodeling in asthma and chronic obstructive pulmonary disease? *Pulm Pharmacol Ther* 19: 18–23

117 Drannik AG, Pouladi MA, Robbins CS, Goncharova SI, Kianpour S, Stampfli MR (2004) Impact of cigarette smoke on clearance and inflammation after Pseudomonas aeruginosa infection. *Am J Respir Crit Care Med* 170: 1164–1171

118 Schols AM, Slangen J, Volovics L, Wouters EF (1998) Weight loss is a reversible factor in the prognosis of chronic obstructive pulmonary disease. *Am J Respir Crit Care Med* 157: 1791–1797

119 Chen H, Vlahos R, Bozinovski B, Jones J, Anderson GP, Morris MJ (2005) Effect of short-term cigarette smoke exposure on body weight, appetite and brain neuropeptide Y in mice. *Neuropsychopharmacology* 30: 713–719

120 Stevenson CS, Koch LG, Britton SL (2006) Aerobic capacity, oxidant stress, and chronic obstructive pulmonary disease – A new take on an old hypothesis. *Pharmacol Ther* 110: 71–82

121 Wright JL, Levy RD, Churg A (2005) Pulmonary hypertension in chronic obstructive pulmonary disease: current theories of pathogenesis and their implications for treatment. *Thorax* 60: 605–609

122 Wright JL, Churg A (1991) Effect of long-term cigarette smoke exposure on pulmonary vascular structure and function in the guinea pig. *Exp Lung Res* 17: 997–1009

123 Zheng T, Zhu Z, Wang Z, Homer RJ, Ma B, Riese RJ Jr, Chapman HA Jr, Shapiro SD, Elias JA (2000) Inducible targeting of IL-13 to the adult lung causes matrix metalloproteinase- and cathepsin-dependent emphysema. *J Clin Invest* 106: 1081–1093

124 Lappalainen U, Whitsett JA, Wert SE, Tichelaar JW, Bry K (2005) Interleukin-1beta

causes pulmonary inflammation, emphysema, and airway remodeling in the adult murine lung. *Am J Respir Cell Mol Biol* 32: 311–318

125 Lanone S, Zheng T, Zhu Z, Liu W, Lee CG, Ma B, Chen Q, Homer RJ, Wang J, Rabach LA et al (2002) Overlapping and enzyme-specific contributions of matrix metalloproteinases-9 and -12 in IL-13-induced inflammation and remodeling. *J Clin Invest* 110: 463–474

126 Underwood DC (1997) Chronic obstructive pulmonary disease. In: DW Morgan, LM Marshall (eds): *In vivo models of inflammation*. Birkhäuser, Basel

127 Nikula KJ, Green FH (2000) Animal models of chronic bronchitis and their relevance to studies of particle-induced disease. *Inhal Toxicol* 12: 123–153

128 Mahadeva R, Shapiro SD (2002) Chronic obstructive pulmonary disease * 3: Experimental animal models of pulmonary emphysema. *Thorax* 57: 908–914

129 Voelkel N, Taraseviciene-Stewart L (2005) Emphysema: an autoimmune vascular disease? *Proc Am Thorac Soc* 2: 23–25

130 Chua F, Gauldie J, Laurent GJ (2005) Pulmonary fibrosis: searching for model answers. *Am J Respir Cell Mol Biol* 33: 9–13

Murine models of allergen-induced airway hyperresponsiveness and inflammation

Azzeddine Dakhama and Erwin W. Gelfand

Division of Cell Biology, Department of Pediatrics, National Jewish Medical and Research Center, 1400 Jackson Street, Denver, Colorado 80206, USA

Introduction

Asthma is a human airway disease characterized by reversible airflow obstruction, persistent airway inflammation and airway hyperresponsiveness (AHR). Both genetically determined host factors and environmental factors are thought to play a role in the expression of the disease in susceptible individuals [1]. Allergic asthma is classically described as being the result of an imbalance in T helper type-1 (Th1)/Th2 immune regulation, resulting in increased production of Th2 cytokines associated with elevated immunoglobulin E (IgE) levels and development of airway eosinophilia. This response is mainly driven by allergen-specific CD4$^+$ T helper cells. The development of Th1 and Th2 responses is complex and may involve many factors including the genetic background [2], the dose and nature of the antigenic stimuli [3], the affinity and strength of peptide binding to MHC molecules [4, 5], the route of sensitization [6], the nature of costimulatory signals and, more critically, the type of cytokines present in the T cell environment during antigen priming and initiation of the T cell response [7]. Circumstances under which humans become sensitized to allergens are not well understood. Apparently, antigen exposure via mucosal surfaces such as the respiratory mucosa favors immune responses associated with Th2 development [6, 8]. At this site, dendritic cells (DCs) appear to play a critical role in the Th1/Th2 differentiation decision. Resting respiratory tract DCs producing low levels of interleukin (IL)-12 are thought to preferentially promote the development of Th2 cells [9], whereas the generation of Th1 cells in the respiratory mucosa may require obligatory cytokine signals, notably IL-12. It is the amount of IL-12, not the cytokine *per se*, that appears to be the most crucial DC factor that drives the development of naïve precursor CD4$^+$ T cells into Th1 cells. DCs in particular can produce large amounts of IL-12 upon encounter with microbial pathogens or their products [10]. In the presence of large amounts of IL-12, the development of Th1 cells is favored because Th2 cells are not responsive to this cytokine due to a selective down-regulation of the IL-12 receptor β2 subunit during their developmental

In Vivo Models of Inflammation, Vol. II, edited by Christopher S. Stevenson, Lisa A. Marshall and Douglas W. Morgan

commitment [11]. Thus, the lack of IL-12 production has been proposed as an alternative "default pathway" that would favor the development of Th2 cells [12]. This pathway is likely to be prevalent at an early age when the immune system is not mature, potentially Th2-biased, and more prone to allergic sensitization [13, 14]. Indeed, early age appears to reflect the most critical window of susceptibility where a sequence of events termed "the atopic march" takes place [15, 16], leading to the staggered development of allergic disease in the predisposed individual.

The last few decades have witnessed significant increases in the prevalence of allergic disorders in Western countries, eliciting several epidemiological studies of the potential factors that could explain such alarming increases. Improvements in public health, changes in lifestyle, and small family size were among early epidemiological observations that led to the hypothesis that early childhood infections may protect against the development of allergies [17]. By analogy, the so-called "hygiene hypothesis" proposed that a lack of exposure to microbial pathogens due to increased hygiene might contribute to the increasing prevalence of allergies in westernized countries. Although providing a plausible explanation, this hypothesis remains controversial and lacks supportive mechanistic insights. The most simplistic explanation, based on the Th1/Th2 imbalance dogma, suggests that microbial pathogens or their products cause an immune deviation favoring the development of counter-regulatory Th1 responses [18]. One such microbial product that received most attention is endotoxin, a lipopolysaccharide shed from the surface of gram-negative bacterial wall. Proponents of the hygiene hypothesis initially argued that early life exposure to endotoxin may deviate the host immune response away from allergic sensitization, potentially interrupting the "atopic march" and subsequent expression of allergic disorders including asthma. Support for this notion is provided by cross-sectional studies, which found an inverse relationship between levels of house dust endotoxin and the frequency of allergic sensitization and atopic asthma, but not with non-atopic asthma, most particularly in rural areas [19, 20]. Other studies, however, found that exposure to high levels of house dust endotoxin was in fact positively associated with wheezing in the first year of life [21], despite a decreased odds ratio for atopic dermatitis at 6 months of age [22]. This raises the possibility that endotoxin exposure, although inversely related to the severity of allergic sensitization, may not prevent the development of asthma. On the other hand, increased hygiene and a presumed lack of microbial infections cannot simply explain the parallel increases seen in the prevalence of Th1-mediated autoimmune diseases such as Type 1 diabetes in Western countries [23]. Perhaps the most striking observations are those made in developing African countries where helminth infections are most common, yet the prevalence of allergic diseases and asthma are the lowest despite evidence of (IgE) sensitization to aero-allergens [24]. These findings appear to contradict the "hygiene hypothesis" as an explanation of the increased prevalence of allergy and asthma. To reconcile these apparently discrepant epidemiological observations, current thinking suggests an alternative explanation to the parallel increases of both Th1- and Th2-mediated diseases in developed countries. The emerging unifying concept is that of an anti-inflammatory regulatory network defined by IL-10 and TGF-β production by antigen-presenting cells or regulatory T cells during microbial and parasitic infections. These cells or their products restrain the development and/or expression of Th2-associated allergic manifestations and organ-specific Th1-mediated autoimmune diseases [24–26].

Although these observations are crucial to our understanding of the disease in humans, the mechanisms underlying the induction and maintenance/progression of allergic airway disorders and asthma remain to be clearly defined. Asthma is a human disease and "man is the best model of human asthma" [27]. However, studies conducted in humans remain limited due to many ethical considerations. To overcome these limitations, experimental animal models are commonly used to test hypotheses and investigate potential mechanisms that may be relevant to understanding the human disease.

Murine models of allergen-induced AHR and inflammation: functional assessments

Asthma is characterized by the presence of inflammation in the airways, which is likely to induce AHR, the cardinal feature of airway dysfunction in asthma. Various animal species have been used as experimental models to mimic aspects of asthma pathophysiology. These models are best described, and should be referred to, as models of allergic airway inflammation and AHR rather than as models of asthma *per se*. Among the species used, mice are most commonly chosen for their several advantages, including well-defined gene and immune functions and the availability of a wide repertoire of gene-modified strains that allow specific definition of the involvement or essential nature of a particular cellular or molecular pathway in the disease process. In addition, assessing lung function in mice has been made possible owing to significant advances achieved in small animal lung physiology [28].

Allergic airway inflammation is most commonly induced in mice by sensitizing these animals to a given allergen and subsequently exposing them to the same allergen via the airways. Airway function, inflammation, and lung histopathology can then be assessed at various time points, both qualitatively and quantitatively. The results may vary depending on the strain of mice and experimental protocol used (i.e., mode and route of sensitization and airway challenge). To some degree, the changes may reflect the heterogeneity of human disease. Sensitization is most often achieved via systemic (subcutaneous or intraperitoneal) administration of the allergen combined with an adjuvant (e.g., alum). This "induction phase" elicits an allergen-specific, Th2-polarized T cell response accompanied by the development of allergen-specific IgE antibodies. To establish allergen-specific responsiveness in the airways, sensitized mice are subsequently exposed to the concordant allergen, which is introduced into the airways by inhalation (as an aerosol), intranasal, or intratra-cheal administration. During this "maintenance phase", inflammatory cells are recruited to the airways followed by the development of AHR. Multiple approach-es can be used to study the sequence of events that lead to the development, main-tenance, and resolution of allergic airway inflammation and AHR in mice. These include: (a) genetic approaches, using gene-modified strains of mice (e.g., gene knockout and overexpression), (b) immunological approaches, using type-specific immune cell depletion or adoptive immune cell transfers, and (c) pharmacological approaches, using drugs and reagents to promote or block certain pathways and functions. These interventions can be tested at various phases of the disease and their impact can be evaluated by several end-point measurements assessing airway function, airway inflammation, and lung histopathology.

Assessment of airway function

AHR is a reflection of lower airway dysfunction and hence direct assessment of lower airway function is critical to understanding the pathophysiology of reactive airway diseases, particularly asthma. Airway function can be assessed in mice both *in vivo*, measuring airway responsiveness to a cholinergic agonist, and *in vitro*, assessing airway smooth muscle function at the pre-junctional, junctional, and post-junctional levels.

In vivo assessments of airway function can be carried out by non-invasive meth-ods using whole body plethysmography [29], measuring Penh (enhanced pause) val-ues in conscious, spontaneously breathing animals, or by invasive methods [30], measuring lung resistance and dynamic compliance in anesthetized and mechanical-ly ventilated animals. The non-invasive methods offer unique opportunities to per-form longitudinal follow-up measurements in the same animals. Respiratory rates and antigen-specific early and late phase responses in addition to nonspecific AHR can be monitored using this method. However, more direct measurements are nec-essary to ensure that airway dysfunction is truly occurring in the lower airways. AHR is most reliably determined by invasive methods directly measuring lung resis-tance and dynamic compliance. AHR is defined as an exaggerated constrictive air-way response to physiological (e.g., acetylcholine: ACh) or pharmacological (e.g., methacholine: MCh) cholinergic agonists that cause airway smooth muscle con-traction. When invasive methods are used, these cholinergic agonists can be admin-istered to the animals systemically, by intravenous injection, or locally, as aerosols generated from an ultrasonic nebulizer and delivered directly to the airways during mechanical ventilation. ACh is the preferred agonist used for challenge by the sys-temic route because it is metabolized faster than MCh, minimizing the effects on other target organs (e.g., heart and vasculature). To some extent these systemic effects may account for the lower sensitivity of AHR detection seen after intra-

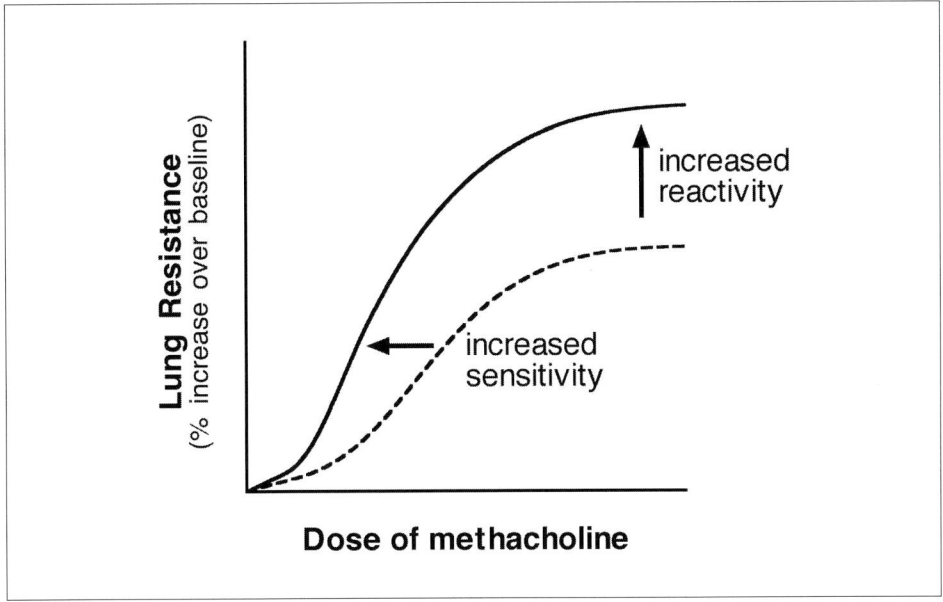

Figure 1
Change in lung resistance in response to methacholine challenge in vivo. Airway hyperre-
sponsiveness is documented by full methacholine dose-response curves reflecting increases
in both airway reactivity (upward shift in the curve) and airway sensitivity (leftward shift in
the curve) to methacholine.

venous challenge as compared with direct airway aerosol challenge. The latter is performed using MCh instead of ACh for stability reasons, allowing enough time for detection and recording of airway responses. When assessing airway responsiveness, full ACh or MCh dose-response curves should be generated to properly document AHR, reflected by increased airway sensitivity and reactivity to the agonist (Fig. 1). These curves can then be used to calculate PC100 or PC200, the provocative concentrations of agonist that cause 100% or 200% increase in lung resistance. Both values can be used to compare airway sensitivity to MCh between groups of animals and treatments.

In vitro assessments consist of measuring contractile responses of isolated mouse trachealis smooth muscle (TSM) segments to pharmacological or electrical field stimulation [31]. These methods further determine if the changes in airway function are transduced at the airway smooth muscle level. In this system, TSM segments (2–3 mm) are mounted in an organ bath containing a physiological solution (e.g., Krebs-Henseleit solution). The segments are supported by two wire supports inserted from the luminal side of the trachea. The lower support is attached to a stainless steel hook at the base of the organ bath, and the upper support is attached to an isometric force transducer connected to a multi-channel amplifier. The transducer is mounted on a rack and pinion clamp permitting adjustments of the resting length of each TSM segment. The baths are aerated and the temperature maintained at 37°C. TSM segments are equilibrated in the bath for 60 min at an optimal resting tension of 1.0 g. During this equilibration time, TSM segments are first challenged with KCl (120 mM) to assess their responsiveness, then rinsed, and allowed to relax to their initial tension. Electrodes are placed in the bath on both sides of the TSM segment to produce an electrical field stimulation (EFS) delivered at increasing frequencies (0.5–20 Hz). EFS stimulates airway cholinergic nerve fibers causing ACh release, which mediates smooth muscle contraction. A sigmoid dose-response curve is obtained by plotting the generated TSM contractions (tension) against the corresponding frequencies of EFS (Fig. 2). For comparisons between the groups, the data are normalized by expressing them as g tension/g tissue or as percent of maximal tension (Tmax). A leftward shift in the curve toward lower frequencies indicates increased TSM sensitivity to EFS. This shift can be reflected by a decrease in the values of ES_{50}, the frequency of electrical stimulation that causes 50% of maximal tension. Such decreases identify changes in airway smooth muscle function occurring at the pre-junctional level upstream of ACh receptors present on the smooth muscle. To detect post-junctional changes in airway smooth muscle function, TSM segments are stimulated with increasing doses of MCh or carbachol, a more stable cholinergic agonist. Similar to ACh, both MCh and carbachol bind to the high affinity muscarinic ACh receptor 3 (M3 receptor) on the surface of airway smooth muscle and cause contraction. A dose-response curve plotting TSM tension values against doses of MCh can be generated. An EC_{50} value, the concentration of MCh causing 50% of maximal tension, can be derived from the plot. A leftward shift in the MCh dose-response curve, also reflected by a decrease in EC_{50} values, indicates that changes in airway function are occurring at the post-junctional level, on airway smooth muscle. Finally, changes in airway function may also occur at the junctional level between nerve and airway smooth muscle. At this level, acetylcholinesterase activity may play a significant role, regulating the levels and thus availability of ACh released from cholinergic nerve fibers.

Assessment of airway inflammation

Airway inflammation can be assessed in many ways using cytological, histological, immunological, molecular, and biochemical approaches. The combination of some or all these approaches may be necessary for defining the mechanisms underlying development of allergen-induced airway inflammation. Inflammatory cells are commonly recovered from the bronchoalveolar space of mice by lavage, instilling an aliquot of balanced physiological salt solution and gently recovering the fluid. The

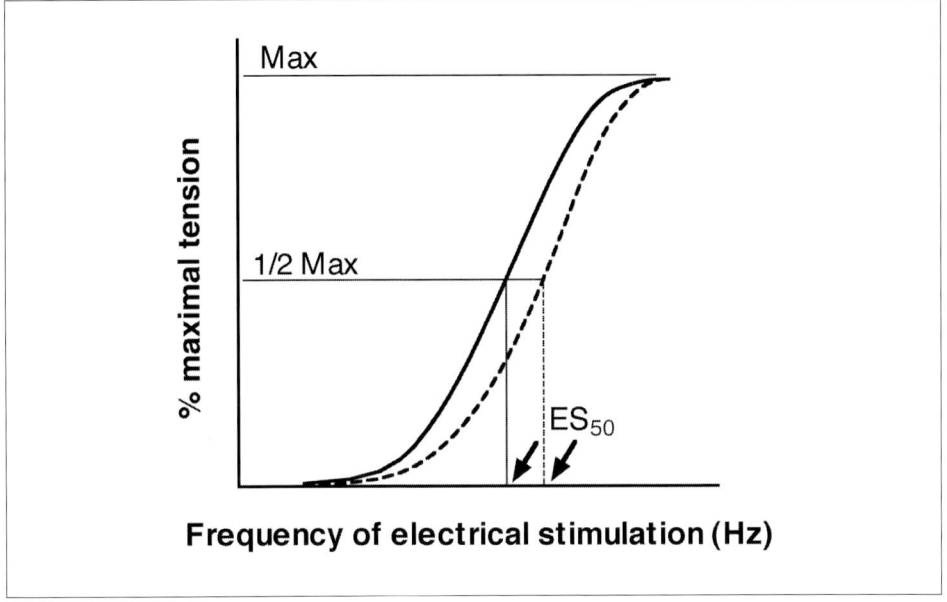

Figure 2
Contractile response of isolated airway smooth muscle segments to electrical field stimulation in vitro. Electrical field stimulation (EFS) elicits a frequency-dependent increase in airway smooth muscle (ASM) contraction caused by release of endogenous acetylcholine from stimulated airway cholinergic nerve fibers. A full EFS frequency-response curve is generated to document changes in airway responsiveness occurring at the pre-junctional level of ASM. A decrease in ES_{50} value, reflected by a leftward shift in the curve (solid line), indicates increased airway responsiveness to endogenous mediators (mostly acetylcholine) of ASM contraction.

recovered bronchoalveolar lavage fluid (BALF) is centrifuged to sediment cells and to recover the BALF supernatant, which contains various mediators of inflammation. Total and differential counts of cells recovered in BALF are used to estimate the type and degree of cellular airway inflammation. However, since BAL cellularity may not necessarily correlate with, nor predict, the type and distribution of cellular inflammation in airway tissue, careful histological analyses should also be considered. The degree of cellular inflammation in tissue can be estimated using histopathological scoring systems. These systems are usually established by individual laboratories, and are difficult to standardize and remain subjective. On the other hand, objective assessments require counting of individual inflammatory cell types in airway tissue, normalizing the data morphometrically to standard units of tissue area. This difficult task implies the use of specific stains to detect various inflammatory cell types infiltrating the tissue. Although cellular inflammation can be heterogeneous in tissue, histological analyses also help define differences in inflammatory cell distribution at various anatomical sites in the lung tissue (central vs peripheral/parenchymal). Alternatively, lung tissue can be digested with collagenase to release tissue-infiltrating inflammatory cells. This is usually performed after perfusion of the lung vasculature with an EDTA-containing solution to avoid recovery of circulating intravascular inflammatory cells. Once recovered, lung tissue inflammatory cells can be differentiated by staining and counting. Phenotypic and functional analyses can be carried out on these cells for further understanding of the immunopathological features of the response. A repertoire of antibodies is commercially available for use to detect individual cell types or mediators inside or on the surface of a cell. The results can be related to a particular cell type and further used to define immune cell responses.

An important aspect of inflammation is the production of a variety of soluble mediators that can diffuse in tissue and body fluids and regulate cellular activity and function. These include cytokines, chemokines, lipid mediators, bioactive peptides, enzymes, and gases. Various biochemical and immunological assays are now available for specific and sensitive detection and quantification of inflammatory mediators. With respect to airway inflammation, these mediators can be recovered in BALF or directly from tissue after homogenization with or without organic extraction. These mediators can also be recovered from culture supernatants after *in vitro* stimulation of inflammatory cells isolated from the lung or regional lymphoid tissues. With a well-defined, fully sequenced, and characterized mouse genome, molecular approaches are becoming very useful for screening and analysis of gene expression and function as they relate to airway inflammation. Mouse genome-wide expression analysis arrays are commercially available and can be used to identify which genes may be involved in the development, maintenance, or resolution of airway inflammation. The methods are sensitive and specific, and may predict protein expression, although the predictive value may be biased by post-transcriptional gene regulation. Proteomic approaches are being developed and may provide complementary information in the future. The data obtained can be very informative; however, none of these methodologies or approaches can predict post-translational protein function or activity, and there is a paucity of screening tools for assessing lipid mediators and function. Thus, a more global and integrated analysis may be necessary in the future to achieve a more complete understanding of the complexities of airway inflammation.

Assessment of lung histopathology

In addition to inflammatory cellular infiltration, other histopathological changes can occur in association with allergic inflammation in the lung. These changes are morphological (structural) as well as physiological (functional). Structural changes include airway fibrosis characterized by collagen deposition in the subepithelial area, submucosa, within the airway smooth muscle bundles and possibly within the adventitia. Elastin content and structure may also change. Since these changes may have an impact on airway function, quantitative analyses of such structural elements are critical to understanding the pathophysiology of allergic airway disorders including asthma. There are indirect methods for estimating lung tissue collagen synthesis, for example measuring hydroxyproline levels in BALF, and elastin degradation, measuring elastin constituents in urine. However, none of these methods accurately reflects the changes in amount or distribution of these structural elements in airway tissue. Morphometric image analyses of tissue are, therefore, useful for quantitative determinations of structural changes in various lung tissue compartments. The methods can be used reliably not only to monitor changes, but also to evaluate efficacy of therapeutic interventions. Image analysis software versions are available from various commercial sources. There is also a free image analysis software version for Macintosh users, which has been developed at the U.S. National Institutes of Health and is available to the scientific community at the Internet address http://rsb.info.nih.gov/nih-image/.

Using image analysis software, images of lung tissue sections stained for collagen (using Sirius Red) or elastin (using pentachrome) are captured under the microscope with a digital camera and transferred to the computer for morphometric analyses. Images of micrometric scales (1-mm scale with 10-μm subdivisions) are taken with the same camera, on the same microscope and at the same magnification, and used for linear calibration of all measurements. Stained areas are outlined manually and only those pixels belonging to the specific stain are highlighted excluding unstained structures and space included within the same area of interest (Fig. 3). With calibrated images, the highlighted pixels are counted automatically and data are converted to metric units of area. Lung tissue inflation during fixation is essential to open up the airway and alveolar spaces, avoiding collapse of tissue in some areas, which give false impression of tissue compartment thickening. In a number of published studies, this we believe resulted in over- or misinterpretation of "chronic changes" so difficult to achieve in repeatedly allergen-challenged mice [32]. To allow for comparisons between the groups the data should be normalized, e.g., to the perimeter of the corresponding airway, for measurements within the airway wall, or to total tissue area for measurements within the parenchyma. For rigorous comparisons, it is critical to use tissue sections of the same thickness and to stain sections from all of the study groups (including appropriate controls) in the same batch, using the same solutions. It is also important to repeat the measurements on similarly stained consecutive tissue sections cut 50 μm from each other. The data are then averaged for each specimen and the mean values calculated for each group and compared among the different groups for the same anatomic tissue location (central vs peripheral airways or parenchymal tissue). Morphometric image analysis can also

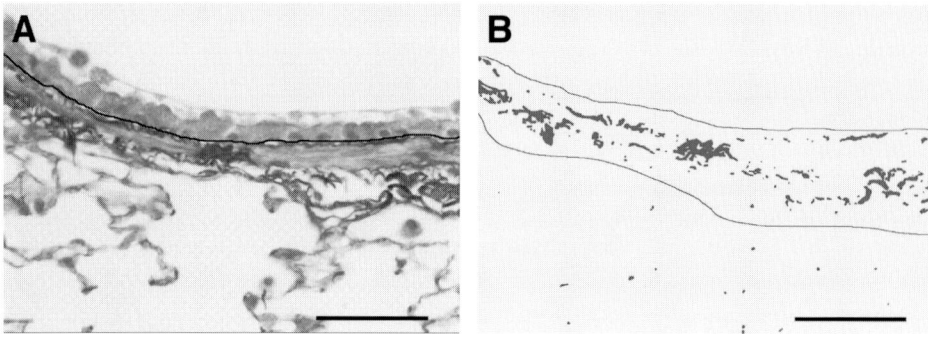

Figure 3
Analysis of airway tissue fibrosis by quantitative morphometry. Digital images of airway tissue sections stained for collagen with Sirius Red are captured under the microscope and transferred to the computer for analysis. Image analysis software (e.g., NIH Image 1.63) is used to measure the perimeter of the basement membrane (BM) of the airway epithelium (A, solid line) and to determine the area of airway fibrosis (B, outlined with solid line), highlighting only those pixels that belong to the stained collagen structures, therefore excluding unstained structures from the measurements. The obtained quantitative data can be expressed as area of collagen (μm^2)/BM perimeter (mm) and compared between groups and treatments. Scale bar represents 50 μm.

be used to quantify the size of other compartments of the airway wall that may play either an active role (e.g., smooth muscle hyperplasia/metaplasia) or a passive role (e.g., epithelial goblet cell metaplasia) in airway narrowing. In addition, certain physiological changes such as intracellular protein expression (e.g., mucus in goblet cells, cytokines or other proteins) can also be quantitatively analyzed by morphometric image analysis, further defining functional changes in different anatomic locations of the lung.

Significance of murine models

The development of allergic airway disease evolves through two important phases: the "induction phase" and the "maintenance phase" (Fig. 4). In the mouse system, these phases are characterized by distinct cytokine and cellular requirements. The induction phase takes place during allergen sensitization and requires IL-4 and CD4 T cells for the initiation of allergen-specific Th2 responses [33, 34]; IL-13 is not required for the development of these responses during the induction phase. The maintenance phase is triggered by airway allergen challenge, which establishes aller-

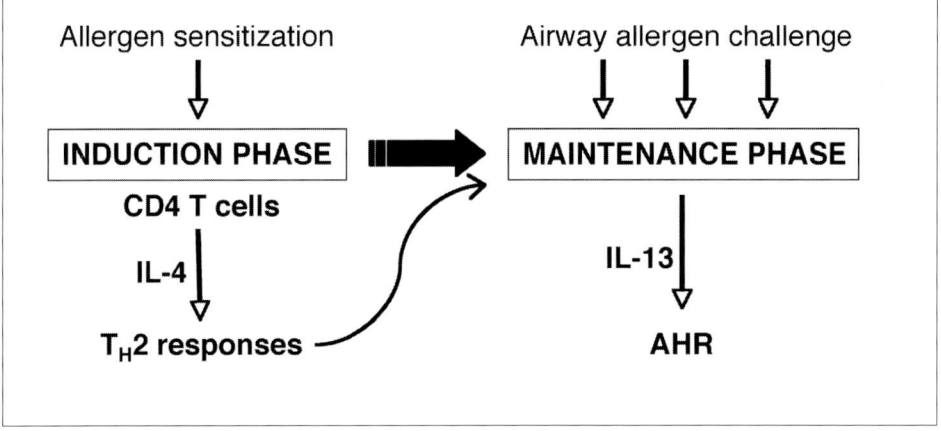

Figure 4
The two-phase model of allergic airway disease development. The development of allergic airway disease consists of two phases. The induction phase takes place during initial sensitization to allergen leading to the development of IL-4-dependent allergen-specific Th2 responses. The maintenance phase is triggered by airway allergen challenge, establishing the response in the airways and leading to the development of airway hyperresponsiveness (AHR) in an IL-13-dependent manner.

gic inflammation in the airways, leading to the development of mucus production and AHR in an IL-13-dependent manner [35–38]; IL-4 is not required for the development of these responses during the challenge (disease maintenance) phase [39, 40]. Similarly, while CD4 T cells are essential for the Th2 responses developing during the induction phase, they are less critical for the development of AHR and lung histopathology during the maintenance phase [41]. Such distinct requirements, and perhaps others that remain to be identified, are likely to influence our development of future therapies for allergic airway disorders, including asthma. Thus, targeting factors that are no longer required for the maintenance of allergic airway disease, several years after the induction phase, would likely not be an effective therapy. This may explain why IL-4 was not as successful as a therapeutic target for the treatment of asthma in adult patients [42]. Nonetheless, IL-4 remains a valid target for intervention during the induction phase to prevent allergic sensitization, which occurs mostly in early childhood. If successful, such intervention may well disrupt the development of the disease in its earliest phase. By contrast, targeting IL-13 appears to be more appropriate for treating established allergic asthma given the involvement of this cytokine in the maintenance phase of the disease. Although it may not reverse the course of the disease or cure it definitively, targeting IL-13 would help attenuate AHR, airway inflammation, and mucus hyperproduction in a way that could prevent the progression of the disease and continuous loss of lung function.

With the use of murine models, several aspects of allergic airway disease became the object of mechanistic investigations. Cause-and-effect rather than associations can be distinguished more clearly. Similar to humans with allergic airway disease, mice can develop both early and late phase allergen-specific airway responses following sensitization and airway allergen challenge [43], providing further opportunity to investigate mechanisms and test novel therapies for allergic asthma. In mice, the early phase response is characterized by a rapid and transient alteration in airway function (increased airway resistance), which develops within 30 min, peaking at around 15 min after airway allergen challenge. This early response is inhibited by treatments with chromoglycate and albuterol. A sustained late phase response follows, developing at around 6 h after airway allergen challenge and can be prevented by treatment with anti-inflammatory steroids. This altered airway response is even more pronounced when previously sensitized and challenged mice are re-challenged with the allergen causing a rapid accumulation and activation of inflammatory cells in airway tissue. In fact, when allergen is administered by inhalation for the first time, it usually takes more than a single aerosol challenge to establish a robust inflammation and AHR in the airways of sensitized mice [44]. Similar to the response elicited by segmental airway allergen challenge in asthma patients, an early but transient neutrophilic airway response can be detected following airway allergen challenge of sensitized mice. This influx of neutrophils is allergen-dependent and requires interaction of allergen-specific IgG antibodies with the low affinity FcγIII receptor [45]. Interestingly, although an allergen-specific late phase response can develop during the same period of time, these neutrophils do not appear to be required for the development of AHR, as depletion of neutrophils during the challenge phase did not prevent the development of AHR in sensitized mice [46].

Previous studies have defined the kinetics for the development of allergen-induced AHR and airway inflammation using a single intranasal allergen challenge model [47]. The data established a parallel between the development of AHR, airway inflammation, Th2 cytokine levels, and mucus production. AHR developed following the decline of the neutrophilic airway response, peaking at 48 h after challenge, and was paralleled by a concomitant accumulation of eosinophils in the airways of sensitized mice. In the BAL, eosinophils appeared at 24 h and their numbers increased to a maximum at 48 h, and were maintained for up to 1 week after challenge. In parallel, tissue eosinophilia reached a maximum by 48 h, but declined progressively, following a similar pattern of resolution as seen for AHR. However, eosinophil activation as indicated by increased eosinophil peroxidase levels in the BALF was detected only at 48 h, coinciding with the peak of AHR. Although lymphocytes appeared in the BALF with a delayed kinetics, their accumulation and activation in tissue likely occurred much earlier, as indicated by increased BAL levels of the Th2 cytokines IL-4 and IL-13 peaking at 24 and 48 h, respectively, after intranasal allergen challenge. Mucus, a perceived mediator of airway obstruction in asthma, did not appear to play a significant role in the development of AHR in this model. In fact, mucus production, as indicated by increased numbers of mucus-containing goblet cells, only reached a maximum by 96 h and was maintained at the same levels for up to 1 week, while AHR began declining during the same period of time. These findings further emphasize the dissociation between mucus production and AHR, and suggest that the quality (type and activation status), not the quantity (amount) of cellular inflammation is what determines the pathophysiology of AHR.

AHR is a complex alteration of airway function for which the mechanisms are not yet fully established. As noted earlier, the route and mode of allergic sensitization can influence the outcome of airway allergen exposure, uncovering different pathways leading to the development of AHR. Specific IgE antibodies are considered to be the best marker of allergen sensitization and their effector role in allergic manifestations is well established, which makes them an important target for immunomodulation of allergic diseases [48]. Cross-linking of mast cell-bound specific IgE antibodies with allergen induces degranulation of these cells and immediate release of inflammatory mediators (histamine), rapid *de novo* synthesis of lipid mediators (leukotrienes) and later transcription and translation of a member of cytokine genes. Varying the protocol of initial sensitization to allergen in mice, two distinct mechanisms of AHR have been identified: IgE-independent/mast cell-independent AHR, and IgE-and mast cell-dependent AHR (Fig. 5). The former develops when mice are systemically sensitized to allergen using alum as adjuvant, eliciting a strong Th2-polarized allergen-specific T cell response. This response is accompanied by a robust allergen-specific IgE response. Surprisingly, neither IgE [49] nor B cells [50] or mast cells [30] are required for the development of allergic airway (eosinophilic) inflammation and AHR following airway allergen challenge in these sensitized mice. These findings suggest that in allergen-sensitized hosts, airway allergen exposure can lead to the development of AHR and airway (eosinophilic) inflammation independent of IgE/mast cell interactions. On the other hand, when repeatedly exposed to allergen exclusively via the airways in the absence of adjuvant and without systemic sensitization, mice develop altered airway function that is best detected by measuring contractile responses of isolated TSM segments to EFS in the organ tissue bath. When compared to control unexposed mice, mice repeatedly exposed to allergen by inhalation develop increased TSM sensitivity to EFS, illustrated by a leftward shift in the EFS frequency-dependent contractile response curve. This altered airway responsiveness is dependent on IgE [51], and is mainly due to increased acetylcholine release from stimulated cholinergic nerve fibers as a result of altered muscarinic acetylcholine M2 receptor function [31]. T cell depletion studies identified CD8 T cells as the major T cell subset mediating such altered airway function [52]. Further investigations into the mechanisms demonstrated that this increased airway responsiveness to EFS requires the high-affinity receptor for IgE

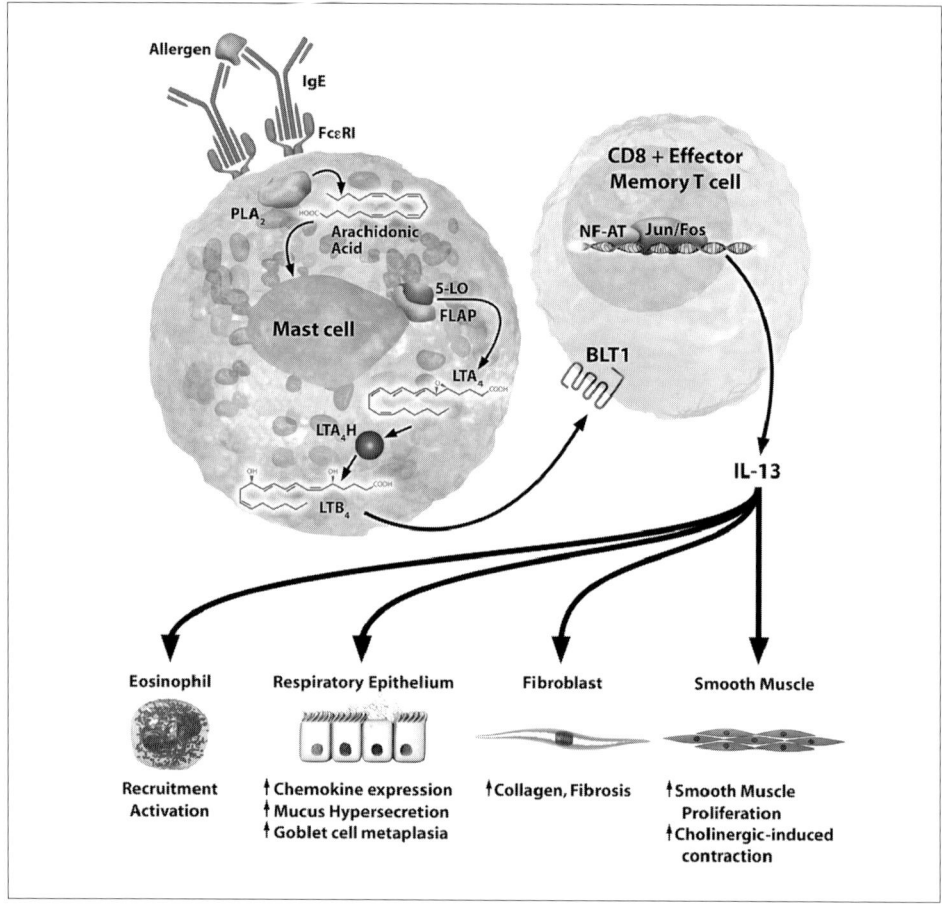

Figure 5

*Involvement of IL-13-producing BLT1⁺ CD8 T cells in the pathogenesis of asthma. Activa-
tion of airway mast cells, by cross-linking of surface-bound IgE with the relevant allergen,
triggers release of LTB4. Binding of LTB4 to its high-affinity receptor, BLT1, on a subset of
CD8 T cells causes their accumulation in the airways. This subset of BLT1⁺ CD8 T cells produces IL-13, which affects many cell types involved in the pathogenesis of asthma. PLA2,
phospholipase A2; FLAP, 5-LO activating protein; LTA4, leukotriene A4; LTA4H, LTA4 hydrolase. (From [90] with permission of the publisher.)*

(FcεRI) and is dependent on mast cells and IL-13 production [53]. Thus, interaction
of allergen with specific IgE bound to FcεRI on the surface of airway mast cells leads
to the development of altered airway responsiveness in an IL-13-dependent manner.
However, although mast cells can be a source of IL-13, the data further suggested

that a cellular source of IL-13 other than mast cells was required for this response. These findings implied that a mediator released from activated IgE-sensitized mast cells might be involved in the response, linking CD8 T cells and IL-13 production in the development of AHR in this allergen challenge model.

Although IL-13 is functionally related to IL-4, sharing a receptor (IL-4Rα) and signaling through the STAT-6 pathway, it nonetheless has distinct and dominant effector roles particularly in allergic airway disease [54]. Over the past few years, IL-13 has emerged as a critical cytokine mediator of allergen-induced AHR [55]. Mice lacking the IL-13 gene or treated with mouse IL-13 receptor α2-human IgG fusion protein (IL-13Rα2-hIgG) fail to develop AHR or mucus metaplasia and have markedly reduced airway eosinophilia following sensitization and airway allergen challenge, establishing the essential role of IL-13 for the development of these altered airway responses [35–38]. Furthermore, transgenic mice overexpressing IL-13 in the lung, under the control of the Clara cell CC-10 promoter, develop marked airway mucus metaplasia, eosinophilic airway inflammation, airway fibrosis, and AHR in the absence of allergen sensitization and airway challenge [56]. These effects of IL-13 appear to be mediated by its direct action on the airway epithelium and require STAT-6 signaling within target airway epithelial cells [57]. Similar observations have been replicated by intratracheal administration of recombinant IL-13 protein in naïve mice [58], further emphasizing the downstream pathogenic effects of this cytokine in allergic airway disease.

Because of their ability to recognize non-self antigens and to produce key cytokines that are involved in allergic inflammation, CD4 T cells have been the central focus in most asthma-related research studies for many years [59]. In the context of allergen sensitization, these cells differentiate predominantly into type-2 cytokine (IL-4, IL-5, IL-9 and IL-13)-producing Th2 cells [60], after receiving the appropriate costimulatory signals and soluble factors, namely IL-4 [34]. Initial studies attributed an important if not dominant role to CD4 T cells in the development of allergen-induced airway inflammation and AHR [61]. In these studies, A/J mice were sensitized by intraperitoneal administration of sheep red blood cells (SRBC) and challenged by intratracheal instillation of SRBC, which resulted in the development of eosinophilic airway inflammation and AHR. Depletion of CD4 T cells before airway challenge, but after sensitization, completely prevented the development of AHR and airway eosinophilia in these studies, implying an essential role for CD4 T cells in the development of these altered airway responses during the disease maintenance (challenge) phase. However, other studies using a murine model of ovalbumin (OVA)-induced airway inflammation and AHR demonstrated that these altered airway responses are fully dependent on T cells, but only partially dependent on CD4 T cells [62], suggesting that a non-CD4 T cell subpopulation could be involved. In addition to Th2-polarized CD4 T cells, other T cells are capable of producing IL-13. These include CD8 (Tc2) T cells [63] and NK-T cells [64].

NK-T cells recognize lipids, lipopeptides and glycolipids that are presented by a family of MHC class I-related non-polymorphic glycoproteins called CD1 molecules [65, 66]. There are five CD1 isoforms (CD1a to CD1e) in human; the mouse locus has a single but duplicated (CD1d1 and CD1d2) CD1 isoform [67]. NK-T cells appear to play important roles in bacterial infections and autoimmunity [68]. However, their role in allergic airway disease is not well understood and still somewhat controversial. NK-T cells are notable for their ability to rapidly produce large amounts of IL-4 and IFN-γ upon T cell receptor stimulation [69, 70], an activity that enables them to function as immunoregulatory cells. Because of this, NK-T cells have initially been thought to be the primary source of IL-4 needed for the initiation of Th2 responses. However, some studies demonstrated that Th2 responses could be initiated in the absence of CD1-dependent NK-T cells [71]. Regardless of the initial source of IL-4, a recent study showed that, unlike wild-type mice, CD1d1 knockout mice and Jα281 (also called Jα18) knockout mice, which lack Vα14 invariant NK-T cells, do not develop AHR and show reduced eosinophilic airway inflammation after sensitization and airway challenge with allergen [64]. Using IL-4- and/or IL-13-deficient donor mice, the study further suggested that invariant NK-T cells producing both IL-4 and IL-13 were required for the development of allergen-induced AHR in sensitized mice. Some of these data were duplicated in another study, suggesting that NK-T cells are required for allergen-induced airway inflammation and AHR [72], although the latter was not properly documented by methacholine responsiveness.

Using α-galactosylceramide (α-GalCer), a marine sponge glycolipid known to activate NK-T cells in mice, a recent study suggested that such an NK-T cell ligand may act as an adjuvant when co-administered with allergen, driving Th2-dependent allergic airway responsiveness presumably by activating NK-T cells [73]. However, recent data demonstrated that a single intraperitoneal or intranasal administration of α-GalCer to allergen-sensitized mice, before airway allergen challenge, completely inhibited the development of airway eosinophilia, lung Th2 cytokine production and mucus goblet cell metaplasia [74, 75]. These inhibitory effects of α-GalCer were attributed to IFN-γ production by activated NK-T cells. More recently, it was shown that intranasal administration of α-GalCer alone, in absence of allergic sensitization, was sufficient for the induction of AHR independent of adaptive immunity, through a mechanism that is dependent on NK-T cells and requires both IL-4 and IL-13 [76]. However, in the same study AHR did not develop when α-GalCer was re-administered again to the mice by the same intranasal route, a result that could be explained by the previously described effect of α-GalCer in inducing NK-T cell anergy [77]. This raises the concern that NK-T cells may not be, as previously anticipated [64], a dominant effector of AHR and asthma in human. After sensitization and airway allergen challenge, the proportion of NK-T cells reaches at most 4% of total lung (Vβ+) T cells. However, at least in the mouse model system, IL-13 is sufficient for inducing AHR in the absence of cellular inflammation [58], and it remains unclear how NK-T cells become activated to release the critical amounts of IL-13 that are required for the induction of AHR, and whether these cells influence the function of other effector cells (i.e., CD4 and CD8 T cells) which are a dominant source of IL-13.

CD8 T cells recognize antigen presented in the context of MHC class I molecules. Similar to CD4 T cells, CD8 T cells can also differentiate into type-1 and type-2 cytokine-producing cells [78]. Consistent with earlier studies, suggesting a role for CD8 T cells in allergen-induced IgE- and mast cell-dependent AHR [52], recent studies of CD8-deficient mice demonstrated that CD8 T cells are required for the development of maximal airway inflammation and AHR in mast cell- and IgE-independent models of allergen-induced AHR [63]. Thus, compared to wild-type mice, systemically sensitized CD8-deficient mice developed significantly lower AHR, reduced airway eosinophilia and mucus production, and lower levels of IL-13 in the BALF after airway allergen challenge. Adoptive transfer of CD8 T cells from sensitized wild-type mice into sensitized CD8-deficient mice restored maximal AHR, IL-13 levels, mucus production, and airway eosinophilia. Transfer of CD8 T cells from naïve wild-type mice or from allergen-sensitized IL-13-deficient mice failed to restore these responses, further demonstrating that CD8 T cells must be primed and produce IL-13 to mediate these altered airway responses [63]. Staining of lung and peribronchial lymph nodes revealed that these cells were indeed a source of IL-13 *in vivo*, indicating that AHR was associated with IL-13 production by primed CD8 T cells.

Antigen-primed CD8 T cells can develop into two subsets of memory cells: central memory (T_{CM}) and effector memory (T_{EFF}) CD8 T cells. These memory CD8 T cell subsets can be distinguished by surface marker expression and tissue-specific homing. T_{CM} home preferentially to lymph nodes and are CD62L^{hi} CCR7^{hi} memory CD8 T cells, whereas T_{EFF} which migrate more efficiently to sites of tissue inflammation are CD62L^{low} CCR7^{low} memory CD8 T cells [79, 80]. T_{EFF} and T_{CM} can be generated *in vitro* by culture with IL-2 and IL-15, respectively. When adoptively transferred into sensitized CD8-deficient mice, prior to airway allergen challenge, T_{EFF} but not T_{CM} migrated to the lung, produced IL-13, and restored maximal AHR, airway eosinophilia, and BALF IL-13 levels [81]. Another important functional characteristic of T_{EFF} is cell surface expression of BLT1, a G protein-coupled high-affinity receptor for leukotriene (LT) B4 [82, 83]. LTB4 is a leukocyte chemoattractant lipid mediator, which is generated during inflammation from cell membrane phospholipids under the sequential actions of cytosolic phospholipase A_2, 5-lipoxygenase and LTA4 hydrolase [84]. Compared to wild-type mice, allergen-sensitized BLT1$^{-/-}$ mice developed lower airway eosinophilia and had lower numbers of IL-13-producing CD4$^+$ and CD8$^+$ T cells in the lung, correlating with lower BALF IL-13 levels and lower AHR after airway allergen challenge [85]. Adoptive transfer of sensitized, but not naïve, BLT1$^{+/+}$ spleen T cells prior to airway allergen challenge fully restored AHR and BALF IL-13 levels in sensitized BLT1$^{-/-}$

mice, emphasizing the important role of the LTB4-BLT1 axis in the development of allergen-induced AHR.

To further define the role of BLT1 in CD8 T cell-mediated AHR, sensitized CD8-deficient mice received either *in vivo* primed CD8 T cells or *in vitro* generated T_{EFF}, from BLT1$^{+/+}$ or BLT1$^{-/-}$ donor mice, followed by airway allergen challenge and assessment of airway inflammation and AHR [86]. When transferred, both BLT1$^{+/+}$, but not BLT1$^{-/-}$, CD8 T cells and T_{EFF} accumulated in the lung and restored AHR, BAL eosinophilia and IL-13 levels, further establishing the role of BLT1 in allergen-mediated CD8 T cell recruitment and induction of eosinophilic airway inflammation and AHR. Similar findings were obtained using a modified allergen-induced, mast cell-dependent AHR model [53], in which the need for sensitization-derived IgE was compensated for by passive sensitization of mice with a mouse monoclonal OVA-specific IgE antibody [87]. In this system, mice were passively sensitized by intravenous administration of OVA-specific IgE and subsequently exposed by inhalation to aerosolized OVA, which results in the development of a significant AHR detected as increased contractile responses of isolated TSM segments to EFS *in vitro*. Using this mast cell-dependent model, unlike wild-type mice, neither CD8-deficient mice nor BLT1$^{-/-}$ mice developed AHR after passive sensitization and airway allergen challenge, further demonstrating that in addition to mast cells and IgE, both CD8 T cells and BLT1 are necessary for the development of this altered airway response [88]. When LTB4 levels were measured in the BALF, the levels were significantly increased after airway allergen challenge in the BALF of wild-type mice and CD8-deficient mice, but not in the BALF of mast cell-deficient mice or in mice lacking FcεRI, the high-affinity receptor for IgE. This suggested that allergen-mediated IgE cross-linking on surface of mast cells induces LTB4 release into the airways of sensitized mice.

To determine the consequence of IgE-mediated mast cell-derived LTB4 release and define the role of BLT1 on the recruitment of CD8 T cells to the lung, particularly T_{EFF}, and subsequent induction of AHR, CD8-deficient mice were passively sensitized with OVA-specific IgE and reconstituted by adoptive transfer of *in vitro* generated wild-type T_{CM} or either BLT1$^{+/+}$ or BLT1$^{-/-}$ T_{EFF}. When transferred into passively sensitized CD8-deficient mice, T_{CM} did not accumulate in the lung and no AHR was detected after airway allergen challenge. Compared to T_{CM}, T_{EFF} produced larger (fivefold) amounts of IL-13 following activation with anti-CD3 and anti-CD28 antibodies. When adoptively transferred into passively sensitized CD8-deficient mice, BLT1$^{-/-}$ T_{EFF} failed to accumulate in the lung and did not mediate AHR after airway allergen challenge. By contrast, BLT1$^{+/+}$ T_{EFF} accumulated in the lung and mediated AHR, a response that was significantly inhibited by treatment of the recipient mice with the LTB4 receptor antagonist CP105,696 or with the murine IL-13Rα2-hIgG fusion protein.

In humans, it does appear that airway CD8 T cells, not eosinophils, correlate with severity of bronchial asthma and predict the loss of lung function [89], suggesting an important role for this subset of T cells in the maintenance and progression of the disease in man as well. Similar to studies in mice, recent studies in humans have documented the existence of IL-13-producing, BLT1-expressing CD8 T cells in the airways of patients with asthma [90]. Interestingly, the number of these cells was increased in the BALF of steroid-resistant (SR) compared to steroid-sensitive (SS) asthmatics, suggesting that either these cells are refractory to steroid therapy or they may contribute to disease severity or both. When recovered by BAL and activated *in vitro*, alveolar macrophages from SR asthmatics released significantly higher amounts (fivefold) of LTB4 compared to alveolar macrophages from SS asthma patients [90], suggesting that the increased numbers of BLT1-expressing CD8 T cells might well be related to increased local production of LTB4 in the airways of asthmatics. Given the importance of mast cells in human allergic airway disease, including asthma, this newly uncovered pathway involving interactions between LTB 4 and BLT1[+] CD8 T cells is likely to play a significant role in the maintenance and or progression of the disease (Fig. 5). Interrupting these interactions and disrupting this pathogenic link may help modulate the disease and perhaps prevent its progression.

Acknowledgments
This work was supported by National Institutes of Health, Grants HL-36577 and HL-61005 and Environmental Protection Agency Grant R825702.

References

1   Cookson W (1999) The alliance of genes and environment in asthma and allergy. *Nature* 402: B5–11
2   Barnes KC, Marsh DG (1998) The genetics and complexity of allergy and asthma. *Immunol Today* 19: 325–332
3   Hosken NA, Shibuya K, Heath AW, Murphy KM, O'Garra A (1995) The effect of antigen dose on CD4[+] T helper cell phenotype development in a T cell receptor-alpha beta-transgenic model. *J Exp Med* 182: 1579–1584
4   Murray JS, Pfeiffer C, Madri J, Bottomly K (1992) Major histocompatibility complex (MHC) control of CD4 T cell subset activation. II. A single peptide induces either humoral or cell-mediated responses in mice of distinct MHC genotype. *Eur J Immunol* 22: 559–565
5   Pfeiffer C, Stein J, Southwood S, Ketelaar H, Sette A, Bottomly K (1995) Altered peptide ligands can control CD4 T lymphocyte differentiation *in vivo*. *J Exp Med* 181: 1569–1574
6   Constant SL, Lee KS, Bottomly K (2000) Site of antigen delivery can influence T cell

priming: pulmonary environment promotes preferential Th2-type differentiation. *Eur J Immunol* 30: 840–847

7   Mosmann TR, Sad S (1996) The expanding universe of T-cell subsets: Th1, Th2 and more. *Immunol Today* 17: 138–146

8   Bochner BS, Undem BJ, Lichtenstein LM (1994) Immunological aspects of allergic asthma. *Annu Rev Immunol* 12: 295–335

9   Stumbles PA, Thomas JA, Pimm CL, Lee PT, Venaille TJ, Proksch S, Holt PG (1998) Resting respiratory tract dendritic cells preferentially stimulate T helper cell type 2 (Th2) responses and require obligatory cytokine signals for induction of Th1 immunity. *J Exp Med* 188: 2019–2031

10 Trinchieri G (1995) Interleukin-12: a proinflammatory cytokine with immunoregulatory functions that bridge innate resistance and antigen-specific adaptive immunity. *Annu Rev Immunol* 13: 251–276

11 Szabo SJ, Jacobson NG, Dighe AS, Gubler U, Murphy KM (1995) Developmental commitment to the Th2 lineage by extinction of IL-12 signaling. *Immunity* 2: 665–675

12 Moser M, Murphy KM (2000) Dendritic cell regulation of TH1-TH2 development. *Nat Immunol* 1: 199–205

13 Holt PG, Clough JB, Holt BJ, Baron-Hay MJ, Rose AH, Robinson BW, Thomas WR (1992) Genetic 'risk' for atopy is associated with delayed postnatal maturation of T-cell competence. *Clin Exp Allergy* 22: 1093–1099

14 Wegmann TG, Lin H, Guilbert L, Mosmann TR (1993) Bidirectional cytokine interactions in the maternal-fetal relationship: is successful pregnancy a TH2 phenomenon? *Immunol Today* 14: 353–356

15 Spergel JM, Paller AS (2003) Atopic dermatitis and the atopic march. *J Allergy Clin Immunol* 112: S118–127

16 Hahn EL, Bacharier LB (2005) The atopic march: the pattern of allergic disease development in childhood. *Immunol Allergy Clin North Am* 25: 231–246

17 Strachan DP (1989) Hay fever, hygiene, and household size. *BMJ* 299: 1259–1260

18 Romagnani S (1994) Regulation of the development of type 2 T-helper cells in allergy. *Curr Opin Immunol* 6: 838–846

19 von Mutius E, Braun-Fahrlander C, Schierl R, Riedler J, Ehlermann S, Maisch S, Waser M, Nowak D (2000) Exposure to endotoxin or other bacterial components might protect against the development of atopy. *Clin Exp Allergy* 30: 1230–1234

20 Braun-Fahrlander C, Riedler J, Herz U, Eder W, Waser M, Grize L, Maisch S, Carr D, Gerlach F, Bufe A et al (2002) Environmental exposure to endotoxin and its relation to asthma in school-age children. *N Engl J Med* 347: 869–877

21 Park JH, Gold DR, Spiegelman DL, Burge HA, Milton DK (2001) House dust endotoxin and wheeze in the first year of life. *Am J Respir Crit Care Med* 163: 322–328

22 Gehring U, Bolte G, Borte M, Bischof W, Fahlbusch B, Wichmann HE, Heinrich J (2001) Exposure to endotoxin decreases the risk of atopic eczema in infancy: a cohort study. *J Allergy Clin Immunol* 108: 847–854

23   Stene LC, Nafstad P (2001) Relation between occurrence of type 1 diabetes and asthma. *Lancet* 357: 607–608

24   Yazdanbakhsh M, Kremsner PG, van Ree R (2002) Allergy, parasites, and the hygiene hypothesis. *Science* 296: 490–494

25   Wills-Karp M, Santeliz J, Karp CL (2001) The germless theory of allergic disease: revisiting the hygiene hypothesis. *Nat Rev Immunol* 1: 69–75

26   Romagnani S (2004) The increased prevalence of allergy and the hygiene hypothesis: missing immune deviation, reduced immune suppression, or both? *Immunology* 112: 352–363

27   Gelfand EW (2002) Pro: mice are a good model of human airway disease. *Am J Respir Crit Care Med* 166: 5–6; discussion 7–8

28   Irvin CG, Bates JH (2003) Measuring the lung function in the mouse: the challenge of size. *Respir Res* 4: 4

29   Hamelmann E, Schwarze J, Takeda K, Oshiba A, Larsen GL, Irvin CG, Gelfand EW (1997) Noninvasive measurement of airway responsiveness in allergic mice using barometric plethysmography. *Am J Respir Crit Care Med* 156: 766–775

30   Takeda K, Hamelmann E, Joetham A, Shultz LD, Larsen GL, Irvin CG, Gelfand EW (1997) Development of eosinophilic airway inflammation and airway hyperresponsiveness in mast cell-deficient mice. *J Exp Med* 186: 449–454

31   Larsen GL, Fame TM, Renz H, Loader JE, Graves J, Hill M, Gelfand EW (1994) Increased acetylcholine release in tracheas from allergen-exposed IgE-immune mice. *Am J Physiol* 266: L263–270

32   Koya T, Kodama T, Takeda K, Miyahara N, Yang ES, Taube C, Joetham A, Park JW, Dakhama A, Gelfand EW (2006) Importance of myeloid dendritic cells in persistent airway disease after repeated allergen exposure. *Am J Respir Crit Care Med* 173: 45–55

33   Kuhn R, Rajewsky K, Muller W (1991) Generation and analysis of interleukin-4 deficient mice. *Science* 254: 707–710

34   Kopf M, Le Gros G, Bachmann M, Lamers MC, Bluethmann H, Kohler G (1993) Disruption of the murine IL-4 gene blocks Th2 cytokine responses. *Nature* 362: 245–248

35   Grunig G, Warnock M, Wakil AE, Venkayya R, Brombacher F, Rennick DM, Sheppard D, Mohrs M, Donaldson DD, Locksley RM et al (1998) Requirement for IL-13 independently of IL-4 in experimental asthma. *Science* 282: 2261–2263

36   Wills-Karp M, Luyimbazi J, Xu X, Schofield B, Neben TY, Karp CL, Donaldson DD (1998) Interleukin-13: central mediator of allergic asthma. *Science* 282: 2258–2261

37   Walter DM, McIntire JJ, Berry G, McKenzie AN, Donaldson DD, DeKruyff RH, Umetsu DT (2001) Critical role for IL-13 in the development of allergen-induced airway hyperreactivity. *J Immunol* 167: 4668–4675

38   Taube C, Duez C, Cui ZH, Takeda K, Rha YH, Park JW, Balhorn A, Donaldson DD, Dakhama A, Gelfand EW (2002) The role of IL-13 in established allergic airway disease. *J Immunol* 169: 6482–6489

39   Cohn L, Homer RJ, Marinov A, Rankin J, Bottomly K (1997) Induction of airway

mucus production By T helper 2 (Th2) cells: a critical role for interleukin 4 in cell recruitment but not mucus production. *J Exp Med* 186: 1737–1747

40 Cohn L, Tepper JS, Bottomly K (1998) IL-4-independent induction of airway hyperresponsiveness by Th2, but not Th1, cells. *J Immunol* 161: 3813–3816

41 Joetham A, Takeda K, Taube C, Miyahara N, Kanehiro A, Dakhama A, Gelfand EW (2005) Airway hyperresponsiveness in the absence of CD4+ T cells after primary but not secondary challenge. *Am J Respir Cell Mol Biol* 33: 89–96

42 Steinke JW (2004) Anti-interleukin-4 therapy. *Immunol Allergy Clin North Am* 24: 599–614

43 Cieslewicz G, Tomkinson A, Adler A, Duez C, Schwarze J, Takeda K, Larson KA, Lee JJ, Irvin CG, Gelfand EW (1999) The late, but not early, asthmatic response is dependent on IL-5 and correlates with eosinophil infiltration. *J Clin Invest* 104: 301–308

44 Dakhama A, Kanehiro A, Makela MJ, Loader JE, Larsen GL, Gelfand EW (2002) Regulation of airway hyperresponsiveness by calcitonin gene-related peptide in allergen sensitized and challenged mice. *Am J Respir Crit Care Med* 165: 1137–1144

45 Taube C, Dakhama A, Rha YH, Takeda K, Joetham A, Park JW, Balhorn A, Takai T, Poch KR, Nick JA et al (2003) Transient neutrophil infiltration after allergen challenge is dependent on specific antibodies and Fc gamma III receptors. *J Immunol* 170: 4301–4309

46 Taube C, Nick JA, Siegmund B, Duez C, Takeda K, Rha YH, Park JW, Joetham A, Poch K, Dakhama A et al (2004) Inhibition of early airway neutrophilia does not affect development of airway hyperresponsiveness. *Am J Respir Cell Mol Biol* 30: 837–843

47 Tomkinson A, Cieslewicz G, Duez C, Larson KA, Lee JJ, Gelfand EW (2001) Temporal association between airway hyperresponsiveness and airway eosinophilia in ovalbumin-sensitized mice. *Am J Respir Crit Care Med* 163: 721–730

48 Infuhr D, Crameri R, Lamers R, Achatz G (2005) Molecular and cellular targets of anti-IgE antibodies. *Allergy* 60: 977–985

49 Mehlhop PD, van de Rijn M, Goldberg AB, Brewer JP, Kurup VP, Martin TR, Oettgen HC (1997) Allergen-induced bronchial hyperreactivity and eosinophilic inflammation occur in the absence of IgE in a mouse model of asthma. *Proc Natl Acad Sci USA* 94: 1344–1349

50 Hamelmann E, Takeda K, Schwarze J, Vella AT, Irvin CG, Gelfand EW (1999) Development of eosinophilic airway inflammation and airway hyperresponsiveness requires interleukin-5 but not immunoglobulin E or B lymphocytes. *Am J Respir Cell Mol Biol* 21: 480–489

51 Hamelmann E, Tadeda K, Oshiba A, Gelfand EW (1999) Role of IgE in the development of allergic airway inflammation and airway hyperresponsiveness – a murine model. *Allergy* 54: 297–305

52 Hamelmann E, Oshiba A, Paluh J, Bradley K, Loader J, Potter TA, Larsen GL, Gelfand EW (1996) Requirement for CD8+ T cells in the development of airway hyperresponsiveness in a marine model of airway sensitization. *J Exp Med* 183: 1719–1729

53 Taube C, Wei X, Swasey CH, Joetham A, Zarini S, Lively T, Takeda K, Loader J, Miya-

hara N, Kodama T et al (2004) Mast cells, Fc epsilon RI, and IL-13 are required for development of airway hyperresponsiveness after aerosolized allergen exposure in the absence of adjuvant. *J Immunol* 172: 6398–6406

54   Wynn TA (2003) IL-13 effector functions. *Annu Rev Immunol* 21: 425–456

55   Wills-Karp M (2004) Interleukin-13 in asthma pathogenesis. *Immunol Rev* 202: 175–190

56   Zhu Z, Homer RJ, Wang Z, Chen Q, Geba GP, Wang J, Zhang Y, Elias JA (1999) Pulmonary expression of interleukin-13 causes inflammation, mucus hypersecretion, subepithelial fibrosis, physiologic abnormalities, and eotaxin production. *J Clin Invest* 103: 779–788

57   Kuperman DA, Huang X, Koth LL, Chang GH, Dolganov GM, Zhu Z, Elias JA, Sheppard D, Erle DJ (2002) Direct effects of interleukin-13 on epithelial cells cause airway hyperreactivity and mucus overproduction in asthma. *Nat Med* 8: 885–889

58   Venkayya R, Lam M, Willkom M, Grunig G, Corry DB, Erle DJ (2002) The Th2 lymphocyte products IL-4 and IL-13 rapidly induce airway hyperresponsiveness through direct effects on resident airway cells. *Am J Respir Cell Mol Biol* 26: 202–208

59   Kay AB (1992) "Helper" (CD4[+]) T cells and eosinophils in allergy and asthma. *Am Rev Respir Dis* 145: S22–26

60   Romagnani S (2002) Cytokines and chemoattractants in allergic inflammation. *Mol Immunol* 38: 881–885

61   Gavett SH, Chen X, Finkelman F, Wills-Karp M (1994) Depletion of murine CD4[+] T lymphocytes prevents antigen-induced airway hyperreactivity and pulmonary eosinophilia. *Am J Respir Cell Mol Biol* 10: 587–593

62   Haile S, Lefort J, Joseph D, Gounon P, Huerre M, Vargaftig BB (1999) Mucous-cell metaplasia and inflammatory-cell recruitment are dissociated in allergic mice after antibody-and drug-dependent cell depletion in a murine model of asthma. *Am J Respir Cell Mol Biol* 20: 891–902

63   Miyahara N, Takeda K, Kodama T, Joetham A, Taube C, Park JW, Miyahara S, Balhorn A, Dakhama A, Gelfand EW (2004) Contribution of antigen-primed CD8[+] T cells to the development of airway hyperresponsiveness and inflammation is associated with IL-13. *J Immunol* 172: 2549–2558

64   Akbari O, Stock P, Meyer E, Kronenberg M, Sidobre S, Nakayama T, Taniguchi M, Grusby MJ, DeKruyff RH, Umetsu DT (2003) Essential role of NKT cells producing IL-4 and IL-13 in the development of allergen-induced airway hyperreactivity. *Nat Med* 9: 582–588

65   Burdin N, Kronenberg M (1999) CD1-mediated immune responses to glycolipids. *Curr Opin Immunol* 11: 326–331

66   Gumperz JE (2006) The ins and outs of CD1 molecules: bringing lipids under immunological surveillance. *Traffic* 7: 2–13

67   Dascher CC, Brenner MB (2003) Evolutionary constraints on CD1 structure: insights from comparative genomic analysis. *Trends Immunol* 24: 412–418

68   Van Kaer L, Joyce S (2005) Innate immunity: NKT cells in the spotlight. *Curr Biol* 15: R429–431

69   Yoshimoto T, Paul WE (1994) CD4pos, NK1.1pos T cells promptly produce interleukin 4 in response to *in vivo* challenge with anti-CD3. *J Exp Med* 179: 1285–1295

70   Bendelac A, Rivera MN, Park SH, Roark JH (1997) Mouse CD1-specific NK1 T cells: development, specificity, and function. *Annu Rev Immunol* 15: 535–562

71   Smiley ST, Kaplan MH, Grusby MJ (1997) Immunoglobulin E production in the absence of interleukin-4-secreting CD1-dependent cells. *Science* 275: 977–979

72   Lisbonne M, Diem S, de Castro Keller A, Lefort J, Araujo LM, Hachem P, Fourneau JM, Sidobre S, Kronenberg M, Taniguchi M et al (2003) Invariant V alpha 14 NKT cells are required for allergen-induced airway inflammation and hyperreactivity in an experimental asthma model. *J Immunol* 171: 1637–1641

73   Kim JO, Kim DH, Chang WS, Hong C, Park SH, Kim S, Kang CY (2004) Asthma is induced by intranasal coadministration of allergen and natural killer T-cell ligand in a mouse model. *J Allergy Clin Immunol* 114: 1332–1338

74   Hachem P, Lisbonne M, Michel ML, Diem S, Roongapinun S, Lefort J, Marchal G, Herbelin A, Askenase PW, Dy M et al (2005) Alpha-Galactosylceramide-induced iNKT cells suppress experimental allergic asthma in sensitized mice: role of IFN-gamma. *Eur J Immunol* 35: 2793–2802

75   Matsuda H, Suda T, Sato J, Nagata T, Koide Y, Chida K, Nakamura H (2005) Alpha-Galactosylceramide, a ligand of natural killer T cells, inhibits allergic airway inflammation. *Am J Respir Cell Mol Biol* 33: 22–31

76   Meyer EH, Goya S, Akbari O, Berry GJ, Savage PB, Kronenberg M, Nakayama T, Dekruyff RH, Umetsu DT (2006) Glycolipid activation of invariant T cell receptor[+] NK T cells is sufficient to induce airway hyperreactivity independent of conventional CD4[+] T cells. *Proc Natl Acad Sci USA* 103: 2782–2787

77   Parekh VV, Wilson MT, Olivares-Villagomez D, Singh AK, Wu L, Wang CR, Joyce S, Van Kaer L (2005) Glycolipid antigen induces long-term natural killer T cell anergy in mice. *J Clin Invest* 115: 2572–2583

78   Sad S, Marcotte R, Mosmann TR (1995) Cytokine-induced differentiation of precursor mouse CD8[+] T cells into cytotoxic CD8[+] T cells secreting Th1 and Th2 cytokines. *Immunity* 2:271–279

79   Sallusto F, Lenig D, Forster R, Lipp M, Lanzavecchia A (1999) Two subsets of memory T lymphocytes with distinct homing potentials and effector functions. *Nature* 401: 708–712

80   Masopust D, Vezys V, Marzo AL, Lefrancois L (2001) Preferential localization of effector memory cells in nonlymphoid tissue. *Science* 291: 2413–2417

81   Miyahara N, Swanson BJ, Takeda K, Taube C, Miyahara S, Kodama T, Dakhama A, Ott VL, Gelfand EW (2004) Effector CD8[+] T cells mediate inflammation and airway hyperresponsiveness. *Nat Med* 10: 865–869

82   Goodarzi K, Goodarzi M, Tager AM, Luster AD, von Andrian UH (2003) Leukotriene

B4 and BLT1 control cytotoxic effector T cell recruitment to inflamed tissues. *Nat Immunol* 4: 965–973

83 Tager AM, Bromley SK, Medoff BD, Islam SA, Bercury SD, Friedrich EB, Carafone AD, Gerszten RE, Luster AD (2003) Leukotriene B4 receptor BLT1 mediates early effector T cell recruitment. *Nat Immunol* 4: 982–990

84 Lewis RA, Austen KF, Soberman RJ (1990) Leukotrienes and other products of the 5-lipoxygenase pathway. Biochemistry and relation to pathobiology in human diseases. *N Engl J Med* 323: 645–655

85 Miyahara N, Takeda K, Miyahara S, Matsubara S, Koya T, Joetham A, Krishnan E, Dakhama A, Haribabu B, Gelfand EW (2005) Requirement for leukotriene B4 receptor 1 in allergen-induced airway hyperresponsiveness. *Am J Respir Crit Care Med* 172: 161–167

86 Miyahara N, Takeda K, Miyahara S, Taube C, Joetham A, Koya T, Matsubara S, Dakhama A, Tager AM, Luster AD et al (2005) Leukotriene B4 receptor-1 is essential for allergen-mediated recruitment of CD8+ T cells and airway hyperresponsiveness. *J Immunol* 174: 4979–4984

87 Oshiba A, Hamelmann E, Takeda K, Bradley KL, Loader JE, Larsen GL, Gelfand EW (1996) Passive transfer of immediate hypersensitivity and airway hyperresponsiveness by allergen-specific immunoglobulin (Ig) E and IgG1 in mice. *J Clin Invest* 97: 1398–1408

88 Taube C, Miyahara N, Ott V, Swanson B, Takeda K, Loader J, Shultz LD, Tager AM, Luster AD, Dakhama A et al (2006) The Leukotriene B4 receptor (BLT1) is required for effector CD8+ T cell-mediated, mast cell-dependent airway hyperresponsiveness. *J Immunol* 176: 3157–3164

89 van Rensen EL, Sont JK, Evertse CE, Willems LN, Mauad T, Hiemstra PS, Sterk PJ (2005) Bronchial CD8 cell infiltrate and lung function decline in asthma. *Am J Respir Crit Care Med* 172: 837–841

90 Gelfand EW, Dakhama A (2006) CD8(+) T lymphocytes and leukotriene B4: Novel interactions in the persistence and progression of asthma. *J Allergy Clin Immunol* 117: 577–582

Skin inflammatory disorders

Lawrence S. Chan

Department of Dermatology, University of Illinois at Chicago, College of Medicine, 808 South Wood Street, MC624, Chicago, IL 60612, USA

Introduction

There are more than 15 different skin conditions that can be considered as inflammatory diseases that are not in the categories of autoimmune, neoplastic, infectious, hereditary, metabolic, or granulomatous diseases. These inflammatory skin diseases include psoriasis, pityriasis rubra pilaris, pityriasis rosea, parapsoriasis, pityriasis lichenoides, lichen planus, lichen nitidus, erythema multiforme/Stevens Johnson syndrome/toxic epidermal necrolysis, dermatitis herpetiformis, subcorneal pustular dermatosis, perioral dermatitis, allergic contact dermatitis, autosensitization dermatitis, Behcet's disease, and atopic dermatitis [1]. While it is impossible to cover all of these diseases in this chapter, I have attempted to include the major chronic diseases that will impact our society the most. This goal leads me to identify diseases that cause the highest burdens to the society. Fortunately, a recent publication entitled *The Burden of Skin Diseases 2005*, prepared by the Lewin Group, Inc. for the Society for Investigative Dermatology and the American Academy of Dermatology Association, serves this purpose perfectly [2]. One way to determine the relative burden of inflammatory skin diseases in our society is to ask what the relative sum spent on prescription drugs have been for these diseases, since most chronic inflammatory skin diseases do not usually require hospitalization. In this Lewin Group's publication, it was determined that the two most costly chronic inflammatory skin diseases are atopic dermatitis and psoriasis, costing 1,480 and 38 millions of dollars, respectively, in the year of 2004 alone [2]. In fact, atopic dermatitis is on the top of the chart as the most costly of all skin conditions in the year 2004, with regard to the prescription drug [2]. There are also intangible costs due to quality of life impact. Atopic dermatitis is ranked the highest of all skin dermatitis conditions in terms of diminished quality of life with a score of 12.2 on the DLQI (dermatology life quality index, with higher score indicating greater life quality impairment) [2]. Psoriasis is ranked somewhat lower than atopic dermatitis with a DLQI score of 8.8, which by comparison is still much higher in life quality impairment than skin fungal infec-

In Vivo Models of Inflammation, Vol. II, edited by Christopher S. Stevenson, Lisa A. Marshall and Douglas W. Morgan
© 2006 Birkhäuser Verlag Basel/Switzerland

tions (DLQI score of 5.5) [2]. With regard to the reason of focusing on chronic type of inflammatory skin diseases, I believe that for the acute type of inflammatory skin diseases, such as allergic contact dermatitis, a straightforward delayed-type hypersensitivity reaction, the major goal should be to identify and avoid the offending agent, the contact allergen. The need for *in vivo* modeling of the acute inflammatory disease such as allergic contact dermatitis, albeit a common one, is lesser in comparison to that of the chronic diseases, in light of the highest value of *in vivo* modeling being the ability to delineate the complex biological responses underlying the inflammation. With the above-stated reasons, I have chosen atopic dermatitis and psoriasis as the focus of this chapter. Due to constrain of the page limitation, only selected models of these two diseases are included in this chapter. Although animal models for these two inflammatory skin diseases have been developed in several animal species, such as mice, rats, and dogs, only mouse models are discussed in here. As one the smallest mammals with rapid reproduction rate, an animal species with a skin structure and an immune system very similar to that of humans [3, 4], an animal species for which abundant genetic information has been delineated, and an animal species for which most of the needed research reagents (such as antibodies to immune components and immune component knockout mice) are readily available [4], mice are the most suitable species for *in vivo* modeling of skin diseases, both for the scientific investigation of pathophysiology of disease and for the pre-clinical therapeutic trial of candidate drugs.

Before discussing the various animal models of these two human diseases, it is appropriate to describe the currently available therapeutic options for these diseases since the ultimate goal of animal modeling is to improve therapeutic modalities for human patients suffering from these diseases. Atopic dermatitis is currently treated predominantly by a variety of topical medications [5, 6]. The most commonly used of these topical medications is corticosteroid, but its long-term use may lead to many undesirable side effects [5, 6]. Topical calcineurin inhibitors (tacrolimus and pimecrolimus) have shown some good therapeutic results in some studies [5, 6]. However, the recent black box inclusion in the medication label, due to a few cases of malignancy that occurred in patients receiving these medications, has cast a big cloud of doubt on the future of these topical calcineurin inhibitors [5, 6]. The treatment for psoriasis is currently tailored according to the severity of the disease [7]. For a local manifestation of psoriasis, i.e., small skin surface involvement, the current treatment options include solo application or a combination of topical corticosteroid, topical tar, topical vitamin D-like compound (calcipotriene), topical anthralin, and topical retinoic acid (tazarotene) [7–9]. For patients with moderate amount of skin involvement, ultraviolet light (narrow band UVB or Psoralen plus UVA), used as a solo treatment or in combination with topical medications, is commonly given [10, 11]. For patients with extensive amount of skin involvement, systemically administered medications, such as retinoic acid (acitretin) or immunosuppressants (methotrexate, cyclosporine), are the choices of the medications [12–14].

Sometimes, these systemic medications are used in combination with ultraviolet light or topical medications. Most recently, systemically administered (intravenous or subcutaneous) biologicals such as TNF-α inhibitors (infliximab, etanercept, and adalimumab), and T cell-targeted therapies (efalizumab and alefacept) have been used with promising results in some studies [15, 16] However, their potential side effects have not yet been thoroughly determined [17].

Atopic dermatitis models

Human atopic dermatitis is defined as a chronic, pruritic (itchy), inflammatory skin disease that predominantly affects children, but can also have adult onset [18, 19]. It is a common skin condition that affects about 20% children and its incidence is on the rise [18]. When the disease affects infants, the lesions, which usually have ill-defined borders, are commonly located on the face and extensor skin surfaces. For toddlers who crawl and walk, the lesions are characteristically located in extensor surfaces. Later in childhood and in adulthood, the most common lesional locations become the flexural areas [18]. Other findings associated with the disease include xerosis (dry skin), staphylococcal skin infection, peripheral eosinophilia, elevation of serum IgE level, ichthyosis, white dermatographism, anterior subcapsular cataracts, keratoconus, and conjunctivitis [18]. Skin barrier abnormalities have also been documented in these patients [18]. Food allergy can be demonstrated in about 40% of patients [18]. Because other mild inflammatory skin conditions can resemble atopic dermatitis [18], some experts in the field have tried to establish diagnostic criteria, so that the accurate incidence and prevalence of the disease can be documented for research purposes [20]. According to one such guideline, the diagnosis of atopic dermatitis can be established if a patient demonstrates a minimum of three major diagnostic criteria plus a minimum of three minor diagnostic criteria [20] (Tab. 1). More recently, other experts in the field of atopic dermatitis research have proposed that atopic dermatitis is not a single disease entity, but rather an aggregation of several similar clinical diseases with certain clinical characteristics in common [21]. They argued that the categorization of atopic dermatitis into an allergic (IgE-associated) and a non-allergic (non-IgE-associated) atopic eczema/dermatitis syndrome at each stage of life (infancy, childhood, teenage, and adult) would be important for the management of these patients together with avoidance of allergens and prevention of secondary allergy [21]. These authors also pointed out that the "atopy march", a term applied for atopic dermatitis patients who go on to develop atopy symptoms (asthma and allergic rhinitis), is much lower in children with the non-IgE-associated type of disease [21]. Furthermore, other experts also suggest that since about two thirds of patients diagnosed to have atopic dermatitis are not "atopic" (without IgE sensitization) further studies are needed to delineate the role of IgE sensitization in atopic dermatitis, and that the term "atopic dermati-

Table 1 - Diagnostic criteria of atopic dermatitis proposed by Hanifin and Rajka [20]

Major criteria:
(minimum 3)
	Pruritus (itchiness)
	Classic clinical morphology and distribution:
	    Adults: flexural lichenification or linearity
	    Infants/children: facial and extensor lesions
	Chronic or chronically relapsing dermatitis
	Personal or family history of atopy (asthma, allergic rhinitis,
	    atopic dermatitis)

Minor criteria: (minimum 3)	Xerosis (dry skin) Ichthyosis/palmar hyperlinearity/keratosis pilaris Immediate (type I) skin test reactivity Elevation of serum IgE Early onset age Prone for skin infection (*S. aureus* & *herpes simplex*)/ impaired cellular immunity Prone for hand or foot dermatitis Nipple eczema Cheilitis Recurrent conjunctivitis Dennie-Morgan infra-orbital line Keratoconus Anterior subcapsular cataracts Orbital darkening Facial pallor/facial erythema *Pityriasis alba* Anterior neck folds Pruritus associated with sweating Intolerance to wool and lipid solvents Perifollicular accentuation Food intolerance Disease course influenced by environmental/emotional factors White dermographism/delayed blanch

tis" is problematically inaccurate [22]. These two subtypes of atopic dermatitis are referred to as "extrinsic" (IgE-associated) and "intrinsic" (non-IgE-associated) by some experts in the field [18, 23]. The strongly proposed reclassification of the two subsets of "atopic dermatitis" not withstanding, there still is no general consensus on this issue at the present time. A world-wide consensus on the definition and diagnostic criteria for atopic dermatitis is very much needed. Nevertheless, when discussing the following animal models of "atopic dermatitis", this potential dichotomy needs to be kept in mind.

Many animal models have been established for the purpose of investigating the pathomechanisms of atopic dermatitis [24]. It is not possible here to cover all of these models in detail. The models described here have been selected to cover a broad range of model systems including both IgE-associated and non-IgE-associated models. Table 2 depicts the comparison of features characteristic of human atopic dermatitis among the mouse models of atopic dermatitis described in this chapter.

NC/Nga spontaneous mouse model

In a publication in 1997, Matsuda et al. [25] reported that the NC/Nga strain of mice, when housed in a conventional (pathogen-present) environment, but not in a special pathogen-free environment, developed an inflammatory pruritic skin condition resembling that of human atopic dermatitis. Skin lesions were located on the face, neck, ears, and dorsal skin, and their severity was correlated with an increase in the levels of serum IgE, like those in human atopic dermatitis [18, 26]. Histopathology and immunopathology of the skin lesions, in comparison to non-diseased skin of the mice housed in a special pathogen-free environment, showed findings typically observed in human atopic dermatitis [18, 27]: acanthosis, spongiosis, hyperkeratosis, parakeratosis, increase infiltration of mast cells, eosinophils, macrophages, CD4+ T cells, and to a lesser extent CD8+ T cells [27]. The skin lesions of these conventionally housed NC/Nga mice also demonstrated an increase of CD4+ T cells that produced IL-4, IL-5, and to a lesser extent IFN-γ [27], similar to that found in human atopic dermatitis [18, 28–30]. Subsequently, overexpression of Th2-specific chemokines TARC (thymus- and activation-regulated chemokine) and macrophage-derived chemokine were demonstrated in the affected mice housed in conventional environment [31]. Furthermore, the etiology of IgE hyperproduction in these NC/Nga mice was found to be through enhanced tyrosine phosphorylation of Janus kinase 3 [32]. This phenomenon could be the result of a higher sensitivity of NC/Nga B cells to CD40 ligand and IL-4, due at least in part to the constitutive phosphorylation of Janus kinase 3 [32], an observation also found in human atopic dermatitis patients [32]. Other observations of this model include impairment of skin barrier functions, as shown by reduction of skin ceramide and increase of transepidermal water loss [33], severe allergic response to intranasal challenge of protein antigen like that of human asthma patients [34], and development of skin lesions upon topical application of mite antigens in a special pathogen-free environment [35]. The skin lesions in these NC/Nga mice responded to topical corticosteroid treatment [36]. Together, the NC/Nga mouse model is clearly an IgE-associated model with a tendency towards "atopy march". The advantage of this

Table 2 - Comparison in mouse models of atopic dermatitis on features characteristic of human atopic dermatitis

Features	NC/Nga strain	Skin allergen-induced	Oral allergen-induced	IL-4-transgenic
Clinical:				
Pruritus[a]	+	+	+	+
Chronic lesions	+	ND	+	+
Lesions@SPF[b]	+	ND	ND	+
Conjunctivitis	−	−	−	+
Familial	+	−	−	+
S.A. infection[c]	−	−	−	+
Xerosis	+	ND	+	+
Res. to TCS[d]	+	ND	+	+
Pathological:				
Spongiosis	+	+	+	+
Acanthosis	+	+	+	+
Hyperkeratosis	+	−	−	+
Parakeratosis	+	−	−	+
MC infiltrate ↑[e]	+	+	+	+
Eo infiltrate ↑[f]	+	+	+	+
Macrophage ↑	+	+	ND	ND
CD4 infiltrate ↑[g]	+	+	+	+
CD8 infiltrate ↑[h]	+	−	+	+
Immunological:				
Serum IgE ↑[i]	+	+	+ antigen-specific	+
Th2 cytokines ↑	+	+	+	+ early
Th1 cytokines ↑	+	+	+	+ late
Non-Th cytokines ↑	ND[l]	−	ND	IL-1β, TNF-α
T cell activation[j]	ND	ND	+	+
B cell activation[k]	ND	ND	ND	+

[a] *Itchiness documented by scratch behavior*
[b] *Skin lesions developed under special pathogen-free environment*
[c] *S.A., Staphylococcus aureus*
[d] *Res., response; TCS, topical corticosteroid*
[e] *MC, mast cells in dermis*
[f] *Eo, eosinophils in dermis*
[g] *CD4, CD4+ T cell subset in dermis*
[h] *CD8, CD8+ T cell subset in dermis*
[i] *Total serum IgE level unless otherwise specified*
[j] *T cell activation ↑ documented by T cell proliferation assay*
[k] *B cell activation ↑ documented by B cell proliferation assay*
[l] *ND, not determined or not documented*

model is that it has been well characterized. The disadvantage of this model is that the NC/Nga mice does not breed well as compared to other common wild-type mice such as BALB/c, C57BL/6, SJL/j, or other outbred strains.

Epicutaneous sensitization mouse model

Wang et al. [37] showed the first time in 1996 that epicutaneous exposure of protein antigen, in the absence of adjuvant, can induce a predominantly Th2 immune response, accompanied by antigen-specific IgE elevation. In 1998 and subsequently 1999, Spergel et al. [38–40] reported that sensitization to protein allergen through the skin of mice can induce a localized inflammatory skin condition, associated with airway hypersensitivity. The epicutaneous sensitization with chicken egg albumin resulted in an elevation of total and antigen-specific IgE and led to development of inflammatory skin lesions at the sensitizing sites [38]. Pruritus was present in the protein-sensitized mice, shown by increased scratching and biting behaviors at the site of sensitization. Histopathological and immunopathological examinations of the inflammatory lesions, in comparison to the normal skin sensitized by saline control, showed acanthosis, spongiosis, and increase inflammatory cell infiltrate including neutrophils, eosinophils, mononuclear cells, mast cells, CD3+ T cells, CD4+ T cell subset, but not CD8+ T cell subset [38]. Semiquantitative analysis of skin lesions, comparing to the normal skin sensitized by saline control, showed increase of IL-4, IL-5, IFN-γ, but not IL-2, TNF-α, and IL-1 [38]. Taking advantage of this skin protein-sensitizing model, these authors subsequently investigated the roles of Th1 and Th2 cytokines by applying the identical sensitizing method in IL-4-, IL-5-, and IFN-γ-deficient mice, and determined that both Th2 and Th1 cytokines play their roles in the inflammatory response of this model, as all three types of cytokine-deficient mice experienced reduction in either epidermal or dermal thickening or inflammatory cell infiltration [39]. Furthermore, dermal inflammatory cell infiltration and IgE elevation were induced by the same protein antigen in TCR δ chain-deficient mice, but not in the TCR α chain-deficient mice, indicating that the αβ T cells, but not the γδ T cells, are essential for the induction of this particular type of allergic skin inflammation [41]. In contrast, dermal mononuclear cell infiltration and IgE elevation were not prohibited when IgH-deficient mice, which lack mature B cells, were sensitized by the protein antigen, suggesting that B cells are probably not critical for the induction of this type of allergic inflammation [41]. Cumulative evidence indicates that this epicutaneous sensitization model is an IgE-associated model [21–23]. The advantage of this model is that it can be generated from wild-type mice at any time the investigators desire, and it does not require any complex induction methods, such as transgenesis, or any sophisticated breeding environment. In addition, many different strains such as BALB/c and C57BL/6 mice can be induced to develop the same inflammatory condition [38, 39]. The disadvantage of this model is that it is primarily a model of acute inflammation, rather than a chronic inflammation, thus significantly limiting its usefulness in determining the immunological sequence of events when a subject's condition is progressively changed from normal to acute disease stage, then to chronic disease stage. Similarly, the testing of candidate drugs in this model may also be restricted to the delineation of effectiveness on acute inflammation, rather than on chronic inflammation.

Oral allergen sensitization mouse model

Li and colleagues [42, 43] reported a mouse model of atopic dermatitis induced by food sensitization. Using C3H/HeJ mice, when intragastrically sensitized to cow's milk or peanut (with cholera toxin as adjuvant), 30–35% animals developed pruritic inflammatory skin lesions, with lichenification, with a tendency to a chronic relapsing disease course. Total skin surface involvement can range from 20% up to 90%. These inflammatory skin lesions responded to topical corticosteroid [42]. Histopathology and immunopathology studies determined that the skin lesions, compared to control normal skin, showed acanthosis, spongiosis, increased dermal infiltration of eosinophils, mast cells, and CD4+ T cells [42]. Antigen-specific IgE elevation and peripheral eosinophilia were also observed [42]. When skin-extracted RNA was used to analyze cytokine profiles by RT-PCR reactions, the food allergen-induced skin lesion, in comparison to non-lesional skin from sensitized mice or naïve mice, showed a substantial increase of IL-5 and IL-13 expression, and a moderate increase of IFN-γ expression [43]. Interestingly, the expression of IL-4 was not increased in the inflammatory skin lesions. This oral allergen-induced mouse model also seems to be an IgE-associated atopic dermatitis model [21–23]. The advantage of this model is that it provides a unique avenue to investigate the mechanism of food-induced atopic dermatitis and it would also provide opportunity to investigate candidate drugs that prevent or manage food-induced atopic dermatitis. The disadvantage of this model is that it represents a small subset of atopic dermatitis, since the majority of patients with atopic dermatitis do not have detectable food allergy [18]. The other shortcoming of this model is that only 30% of sensitized mice developed the inflammatory skin lesions.

IL-4 transgenic mouse model

In 2001, Chan et al. [44] reported the generation of a transgenic mouse model of atopic dermatitis. By overexpressing IL-4 through a keratin 14 promoter/enhancer to the basal epidermis of the keratinized squamous epithelium, the mice spontaneously developed a chronic, inflammatory, skin disorder, in both conventional and special pathogen-free housing environments [44, 45]. Using an improved genotyp-

Figure 1
Clinical phenotypes of atopic dermatitis in the haired IL-4-transgenic mice
Clinical manifestations of atopic dermatitis in the haired IL-4-transgenic mice (right) with a
non-transgenic littermate (left) for comparison. Chronic inflammatory skin lesions are seen
predominantly on relatively hairless skin on the ear and around the eye.

ing method, they also determined that all the IL-4-transgenic mice, but not any of the non-transgenic littermates, developed this inflammatory disorder [45]. Interestingly, they found that the skin disorder occurred predominantly in the relatively hairless skin areas, such as ear (Fig. 1), around the eye, face, and neck, and it occurred only rarely in the haired skin areas, such as back [44]. Histopathologically, the IL-4-transgenic mice started to show changes even before clinical phenotype was observed, with increase of dermal vasculature and mononuclear cell infiltration, in comparison to non-transgenic littermate (Fig. 2a, b). The early skin lesion (1 week old) depicted changes characteristic of acute dermatitis: acanthosis, spongiosis, and increase dermal vasculature and mononuclear cell infiltration (Fig. 2c). The late skin lesion (3 week old) showed changes typical of chronic dermatitis: acanthosis, spongiosis, hyperkeratosis, parakeratosis, and increase dermal inflammatory cell infiltration (Fig. 2d). These findings are consistent with those

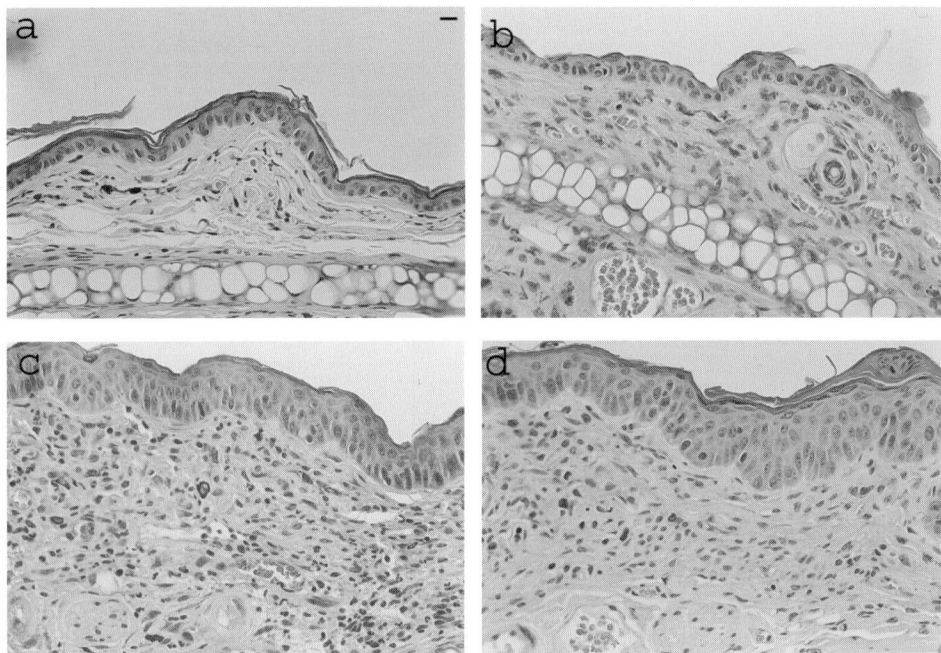

Figure 2
Histopathological findings of skin lesions in the haired IL-4-transgenic mice show changes characteristic of chronic dermatitis like those of human atopic dermatitis
(a). Skin from non-transgenic littermate show an epidermis with two layers of keratinocytes and a dermis with sparse mononuclear cell infiltration. (b). Skin from an IL-4-transgenic mouse before development of clinical phenotype show some degrees of increase of dermal microvasculature and mononuclear cell infiltration. (c). One-week-old skin lesion from an IL-4-transgenic mouse shows substantial changes that are consistent with an acute dermatitis: acanthosis, mild spongiosis, increase dermal vasculature and prominent dermal inflammatory cell infiltration. (d). Three-week-old skin lesion from an IL-4-transgenic mouse depicts pathology consistent with a chronic dermatitis: acanthosis, hyperkeratosis, parakeratosis, spongiosis, and prominent dermal inflammatory cell infiltration. Bar (a–d) = 5 μm, hematoxylin and eosin staining.

found in human atopic dermatitis [18, 27]. When the peripheral and skin-infiltrating leukocytes were examined, it was found that, as the disease progresses from before onset, to early skin lesion to late skin lesion, the T cells were more activated (as shown by their proliferation both spontaneously and in response to stimulants, and by their surface activation molecules) and were more numerous in the skin [46]. In the skin, both CD4+ and CD8+ T cells infiltrated the inflammatory lesions, with CD4+ T cells predominating [46]. When the total RNA extracted from normal appearing skin of non-transgenic littermates and transgenic mice before disease onset, and from skin lesions from transgenic mice with early (1 week old) or late (3 week or older) lesions was examined quantitatively by real-time PCR after reverse transcription, it was found that there was an early upregulation of Th2 cytokines followed by a late surge of Th1 cytokines (Fig. 3) [45], similar to that found in human atopic dermatitis studies [18, 28–30]. Furthermore, most of these IL-4-transgenic mice showed elevation of total serum IgE at the time of disease onset [47], in comparison to their non-transgenic littermates, also similar to that of human atopic dermatitis [18, 26]. The presence of high level of total serum IgE in these diseased mice supports a notion that these mice would be suitable for the IgE-associate atopic dermatitis model [21–23]. The skin inflammation responded to medium-strength topical corticosteroid [48]. In addition, an inflammation-mediated angiogenesis was also observed in this mouse model, documented by characteristic ultrastructure changes in transmission and scanning electron microscopy and confocal microscopy [49]. Although angiogenesis in human atopic dermatitis has not been documented, this observation could help us better understand the relationship between inflammation and angiogenesis, and could potentially lead to a new avenue of anti-inflammatory therapy, as anti-angiogenesis treatment has resulted in suppression of inflammation in animal models of inflammation-mediated angiogenesis [50, 51]. The advantage of this model is that the disease development in the IL-4-transgenic mice is 100%, thus providing the investigators the predictability of disease induction. Since all transgenic mice eventually develop the disease, one can study the progression of the disease, from before onset, to early (acute) skin lesion, to late (chronic) skin lesion, along with various accompanying immunological parameters, so that an immunological sequence of events can be deduced from these studies. Furthermore, mating this line of transgenic mice with mice deficient in immunological components, and observing the disease induction and progression, could help in delineating the role of each of immune components in the induction and maintenance of the disease. With the knowledge of immunological sequence of events for the disease development, target-specific immunological treatments can then be developed to reverse the course of the disease. Similarly, since all transgenic mice will develop the disease, candidate drugs can be tested on this model, not only for their therapeutic effects on the skin lesions already developed, but also for their capacities to prevent the skin lesion development in the first place. The limitation of this model is the continuous presence of IL-4 transgene in the animals throughout their entire life span. Therefore, one cannot answer the question of whether the disease would continue to progress in the absence of IL-4 transgene once the skin inflammation has been initiated. The availability of a tetracycline-controlled transgene expression vector may help to overcome this particular limitation of the model [52].

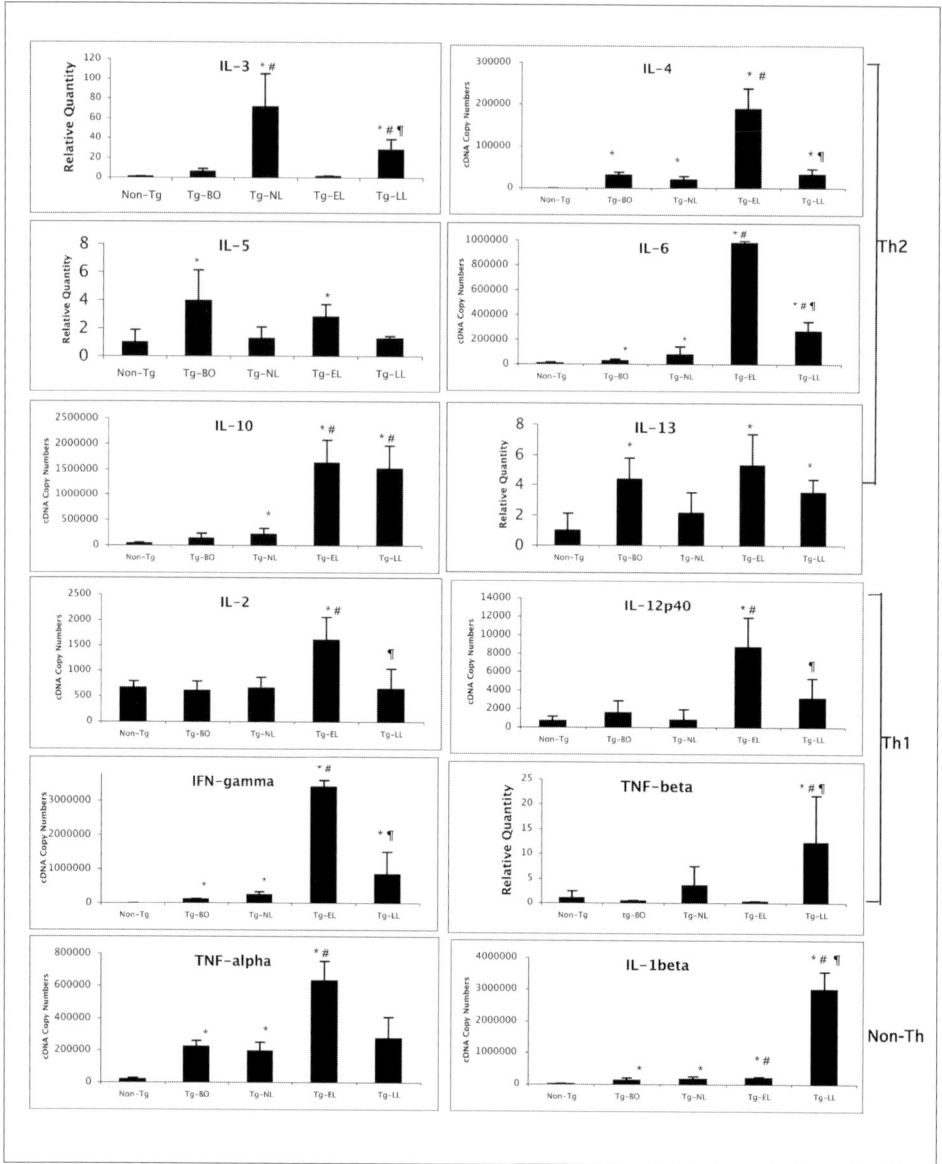

Figure 3
Cytokine profiles of haired IL-4-transgenic mice and their non-transgenic littermates determined by reverse transcription real-time PCR
Statistically significant upregulation of Th2 cytokines (IL-3, IL-4, IL-5, IL-6, IL-10, IL-13) is initially detected either before lesion onset (Tg-BO) or at non-lesional skin (Tg-NL) of the IL-4-transgenic mice, whereas a significant surge of Th1 cytokines is initially detected either

When the IL-4 transgene was transferred from the original haired mice to a hairless but euthymic and immunocompetent mouse strain (SKH1, Charles River) that was housed in special pathogen-free environment, the skin lesions occurred prominently in locations that rarely occurred in the haired mice, such as the back (Fig. 4), suggesting that the hair is a protective factor that acts as a barrier (unpublished publication from the author's laboratory). Further examinations documented that the skin lesions (3 weeks or older) from these hairless mice showed typical chronic dermatitis pattern like that of haired mice (Fig. 2), with acanthosis, spongiosis, hyperkeratosis, parakeratosis, substantial increase of dermal infiltration of mast cells (Fig. 5), eosinophils, and CD4$^+$ and CD8$^+$ T cells (Fig. 6), when compared with that of non-transgenic hairless littermates. In addition, quantitative analyses of cDNAs reverse transcribed from total RNA extracted from the skin of chronic lesions of these hairless mice, in comparison to their non-transgenic hairless littermates, showed substantial upregulation of Th2, Th1, and non-Th pro-inflammatory cytokines: IL-4, IL-6, IL-10, IL-2, IL-12, IFN-γ, and IL-1β (unpublished results from the author's laboratory). The upregulation pattern of cytokines is essentially identical to that in the haired IL-4-transgenic mice (Fig. 3). Interestingly, most of the hairless IL-4-transgenic mice (eight out of ten) that developed inflammatory skin lesions did not have an elevation of total serum IgE (unpublished results). Even in the two of ten hairless IL-4-transgenic mice with skin lesions and elevation of total serum IgE, the levels of IgE were less than 30% of that found in the haired IL-4-transgenic mice with skin lesions (unpublished results). Would this hairless mouse disease model be a suitable non-IgE-associated atopic dermatitis model [21–23]? More in-depth examinations on the IgE-association aspect of this model would certainly help defining its usefulness as a non-IgE-associated atopic dermatitis model. In addition to the potential of being a suitable non-IgE-associated atopic dermatitis model, the hairless IL-4-transgenic model could also be a pre-clinical therapeutic model superior to that of haired IL-4-transgenic model in two aspects. First, the hairless mouse skin, which is structurally closer to human skin (a hairless skin), may be a better medium to test candidate drugs, particularly the topically applied medications. Sec-

---

*at early skin lesions (Tg-EL) (IL-2, IL-12), or late skin lesions (Tg-LL) (TNF-β) of the IL-4-transgenic mice, with the only exception of IFN-γ, which shows a low, but significant elevation before the onset stage. In addition, non-Th pro-inflammatory cytokines IL-1β and TNF-α are also highly upregulated. The quantification was either expressed as relative quantity, with non-transgenic littermate (non-Tg) samples being one, or as cDNA copy numbers calibrated with cytokine plasmid controls. * Statistically significant increase compared to non-transgenic littermates; # significant increase compared to IL-4-transgenic mice before disease onset; ¶ significant difference between early and late skin lesions of the IL-4-transgenic mice (Reprinted from [45], with permission obtained from Blackwell Publishing, Ltd.)*

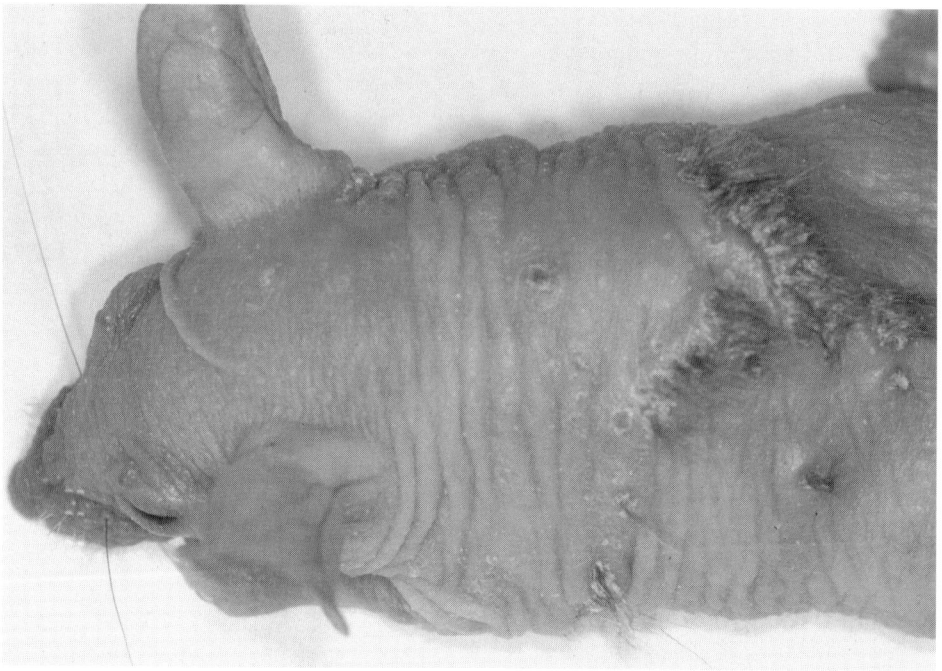

Figure 4
Clinical phenotypes of atopic dermatitis in the SKH1 IL-4-transgenic mice
Clinical manifestations of atopic dermatitis in the hairless IL-4-transgenic mouse, with prominent inflammatory skin lesions on the back.

ond, the larger areas of skin lesions available in the hairless IL-4-transgenic mice, in comparison to the haired mouse counterpart, provide a more optimal condition to apply topical candidate drugs and to observe their therapeutic effects.

Psoriasis models

Unlike atopic dermatitis, a diagnostic criterion or a set of diagnostic criteria for psoriasis that is generally recognized has not yet been adequately addressed. A possible reason is that most experts in the field thought that the diagnosis of psoriasis is straightforward without the need a set of criteria [53]. Thus far, only one article has been published to argue for the need of a set of diagnostic criteria for psoriasis, so that borderline cases can be included or excluded by a more defined set of criteria [53]. Most dermatology textbooks describe psoriasis as a chronic inflammatory skin disease that manifests with well-circumscribed (or demarcated) plaque lesions with

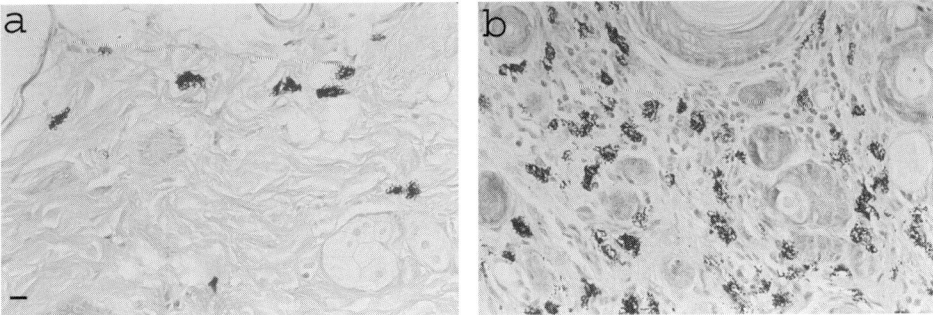

Figure 5
The mast cell infiltration in atopic dermatitis dermis
The dermis of a 3-week-old lesion from a hairless IL-4-transgenic mouse (b), in comparison to that of a hairless non-transgenic littermate (a), illustrates substantial increase of mast cell infiltration. Bar (a, b) = 5 μm, Giemsa staining.

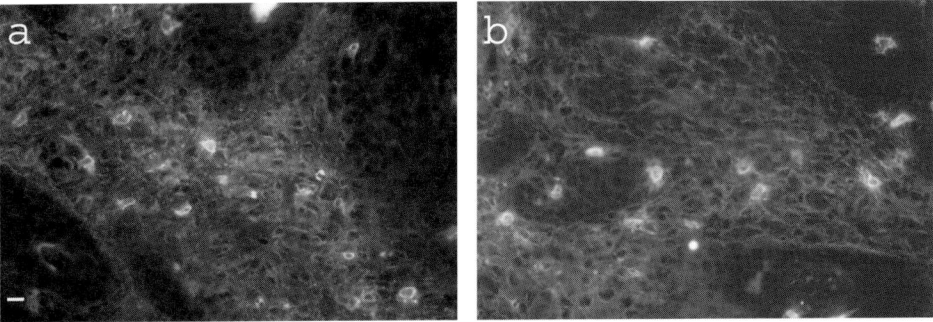

Figure 6
The CD4⁺ and CD8⁺ T cell infiltration in the skin of IL-4-transgenic mice
Immunofluorescence staining by rat monoclonal antibodies to mouse CD4 (a) and CD8 (b), followed by fluorescence (Alexa fluor 488)-labeled goat anti-rat IgG demonstrates the presence of epidermal and dermal infiltrations in the 3-week-old skin lesions of a hairless IL-4-transgenic mouse. Bar (a, b) = 5 μm.

sharply defined border and silvery scales on the surface [54, 55]. The location of the chronic plaque-type lesions, which are the most common form, can either be constrained to the extensor skin surfaces such as elbows, knees, buttocks, intergluteal folds, scalp, and dorsal hands, or be generalized [54, 55]. Koebner's phenomenon, a psoriasis lesion arising from an injured non-lesional skin, is commonly observed

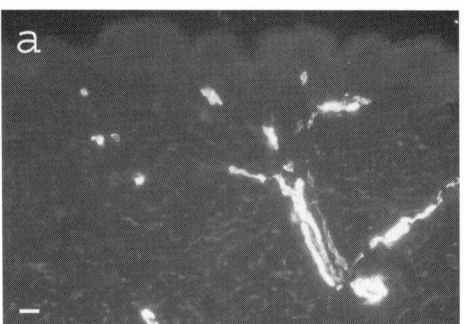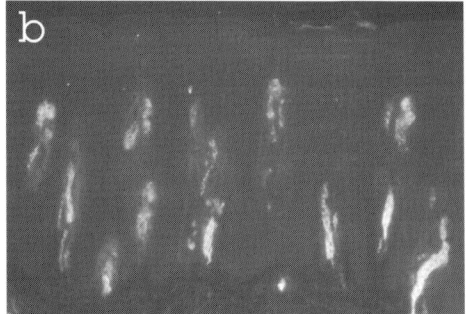

Figure 7
Prominent increase of vasculatures in the human psoriasis lesional skin
Immunofluorescence staining of rabbit anti-human factor VIII-related antigen, followed by
FITC-conjugated goat anti-rabbit IgG labels few dermal blood vessels in a normal human
skin (a) but identifies increases number of dermal blood vessels in a human psoriatic lesion-
al skin (b). Bar (a, b) =10 μm.

in patients with psoriasis [54, 55]. In a small subset of patients, the lesional location is primarily on the flexure areas, opposite of the typical extensor locations, and has been termed "inverse psoriasis" (or psoriasis inversa) [54, 55]. A thinner plaque-type lesion, termed "guttate psoriasis", is an acute type of disease most often affecting young adults, generally recognized as a reaction pattern to streptococcal infection [54, 55]. Sometimes, pustules predominate the clinical phenotype and the disease is termed "pustular psoriasis" [54, 55]. A severe and generalized form, termed "erythrodermic psoriasis" (or psoriatic erythroderma) manifests as diffuse erythema and scaling, affecting the entire body and associated with fever, chills, malaise, fatigue, and pruritus [54, 55]. Psoriasis affects approximately 1–2% of U.S. population [54, 55]. Histopathologically, the chronic plaque lesions show hyperkeratosis, parakeratosis, acanthosis, elongation of rete ridges, loss of granular layer over the tip of dermal papillae, neutrophil microabscesses (Munro's, within the stratum corneum; and Kogoj's, below the stratum corneum) in the upper epidermis, and prominent dermal and epidermal mononuclear cell infiltrate [54]. Characteristically, there are prominent increases of dermal vasculatures (Fig. 7) [54] and T cell infiltrations in both epidermis and dermis (Fig. 8) [54]. On the other hand, markers of epidermal proliferation, including aberrant expressions of keratins and integrins, have also been documented in psoriasis lesional skin [56, 57]. Not surprisingly, there has been a long debate with regard to whether the primarily abnormality of psoriasis is due to epidermal cells or due to the immune system (or cells), and the answer could be provided by animal models of the disease, which were not available until recently.

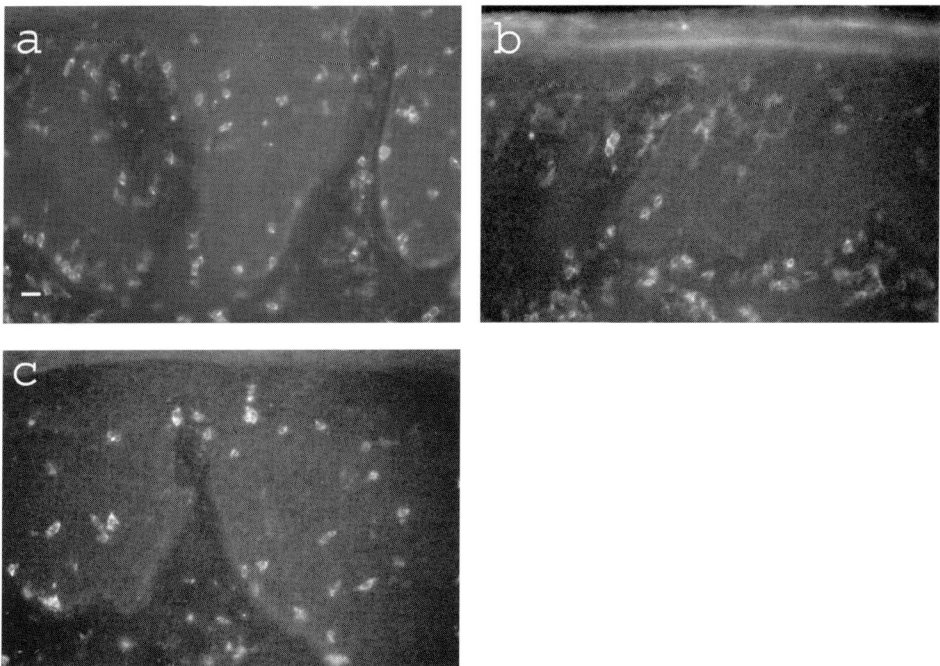

Figure 8
T cell infiltration characterizes the human psoriasis lesional skin
Immunofluorescence staining of monoclonal mouse anti-human pan T cells (a), CD4⁺ T cell
subset (b), and CD8⁺ T cell subset (c), followed by FITC-conjugated goat anti-mouse IgG
illustrates the epidermal and dermal infiltrating T cells and their major subsets. CD4⁺ T cells
are predominantly located in the dermis. Bar (a–c) = 10 μm.

Prior to recent animal model development, investigators used models such as arachidonic acid-induced skin inflammation, phorbal ester-induced skin inflammation, or contact sensitization skin reaction for the purpose of investigating psoriasis. Although these models may provide certain insights into skin inflammation in general, they lack disease specificity, and therefore cannot be recognized as authentic models of psoriasis as such. In particular, these models induced skin inflammation of the acute type, the mechanism of which is very different from that of chronic type of human skin inflammation such as psoriasis. Secondly, although these models are models of skin inflammation, they lack the clinical phenotypes that are characteristic of human disease psoriasis, such as chronicity, well-demarcated plaques with silvery scales, and Auspitz or Koebner signs. Thirdly, these models lack histological features that define human disease psoriasis, such as elongation of rete ridges,

Table 3 - Comparison in mouse models of psoriasis on features characteristic of human psoriasis

Features	Integrin-Tg	HS-SCID mice[a]	VEGF-Tg	TGFβ-Tg	Stat3-Tg	Tie2-Tg
Clinical:						
Plaques[b]	+	+	+	+	+	+
Silvery scales	+	+	+	+	+	+
Pustules	+	ND	ND	ND	ND	ND
Auspitz sign[c]	NDk	ND	ND	ND	ND	+
Koebner sign[d]	ND	ND	+	+	+	+
Pathological:						
Hyperkeratosis	+	+	+	+	+	+
Parakeratosis	+	+	+	+	+	+
Loss granular	+	+	ND	+	+	ND
Acanthosis	+	+	+	+	+	+
Vascularity ↑	+	+	+	+	+	+
Elongation[e]	−	+	+	−	+	+
Munro m.a.[f]	+	+	+	−	+	ND
Kogoj m.a.	+	+	+	+	ND	+[m]
Ki67+ cells ↑	+	+PCNA[l]	ND	ND	+	ND
Immunological:						
CD4 infiltrate ↑[g]	+	+	+	+	+	ND
CD8 infiltrate ↑[h]	+	+	+	+	+	ND
Adoptive transfer[i]	ND	+	ND	ND	+	ND
Responsive to ATM[j]	ND	ND	ND	ND	ND	+

[a] HS, human skin; SCID, severe combined immunodeficiency
[b] Well-circumscribed (sharply demarcated) type of plaques with clear boarder
[c] A phenomenon in which bleeding points are revealed when a scale is mechanically removed by scrapping
[d] A phenomenon in which skin lesion develops in a traumatized normal appearing skin, most often in a linear fashion
[e] Elongation of the rete ridges
[f] m.a., microabscesses
[g] CD4, CD4+ T cell subset
[h] CD8, CD8+ T cell subset
[i] Transferring of immunocytes (such as T cells) from the diseased animal (or human in case of HS-SCID mouse model) to the recipient naïve animal resulted in clinical phenotype development
[j] ATM, anti-T cell medication
[k] ND, not determined or not documented
[l] PCNA, proliferating cell nuclear antigen
[m] Neutrophil microabscesses in the lower epidermis

increase of dermal microvascular vascularity, and Munro and Kogoj microab-scesscss. For these three major reasons, these prior models are no longer used as models for psoriasis, as they would not yield a disease-specific pathogenic mecha-nism, a necessary step for the development of disease-directed target-specific treat-ment.

Before the integrin-transgenic mouse model [58] and the human skin-SCID mouse chimeric model [59] were generated, there were many animal models show-ing some degrees of resemblance to psoriasis. However, these two models are the first to show a clearer picture of clinical and histopathological psoriasis [58, 59]. In particular, the human skin-SCID mouse chimeric model was the first to show the histology of elongation of rete ridges characteristic of lesional psoriasis in human disease [59]. Subsequently, several other transgenic mouse models have been demon-strated to show a good resemblance to human psoriasis and are also discussed below. A comparison of features characteristic of human psoriasis present in these models is summarized in Table 3.

Integrin-transgenic mouse model

In 1995, Carroll et al. [58] reported the generation of a skin disorder with a psori-atic phenotype by transgenically expressing an integrin molecule in the epidermis. The hypothesis for this model was that psoriasis is primarily an epidermal cell dis-order, that initiates the disease, followed by a response of abnormal inflammatory cell infiltration. Epidermal keratinocytes express several extracellular matrix recep-tors for the integrin family. Three of these share a common β1 subunit and are asso-ciated with known functions: α2β1 is a collagen receptor, α3β1 is a laminin recep-tor, and α5β1 is a keratinocyte fibronectin receptor [58, 60]. Under normal condi-tions, complex transcriptional and translational regulatory mechanisms constrain the integrin expression in the basal layer of epidermis [61, 62]. Suprabasal expres-sion of integrin has documented when skin is undergoing a wound healing process or has psoriasis lesions, suggesting a role of suprabasal integrin expression in the induction of psoriasis [57, 58, 60]. By transgenically expressing integrin β1 in suprabasal epidermis using a promoter of involucrin (a suprabasally expressed pro-tein) and by subsequent crossing integrin β1-transgenic mice with integrin α2- or α5-transgenic mice, these investigators showed that aberrant expression of integrin β1 alone in the suprabasal epidermis, or in combination with integrin α2 or α5 (also in the suprabasal epidermis) leads to a psoriasis-like phenotype. This resulted in epi-dermal hyperproliferation, aberrant keratinocyte differentiation, and inflammatory reaction, in addition to some developmental defects [58]. Scaly and inflammatory skin changes were noted in the chin, behind the ears, under the limbs, back, and abdomen skin, with some pustules observed on the back after the transgenic mice reached 6 weeks of age [58]. Histopathology of the lesional skin showed acanthosis, hyperkeratosis, parakeratosis, loss of granular layer underlying the parakeratotic areas, neutrophil microabscesses within (Munro's) and below (Kogoj's) the stratum corneum [58]. In addition, dermal infiltration of mononuclear cells and neutrophils, as well as dilated capillaries, were observed [58]. Immunopathology examinations of the skin lesions revealed markers of keratinocyte hyperproliferation: increased Ki67 expression and aberrant keratin 6 expression in hyperproliferative epidermis [58]. Furthermore, CD3+ T cells infiltrated both epidermis and dermis, with more CD8+ T cells in epidermis and more CD4+ T cells in the dermis [58]. These investigators suggested a possible mechanism in which the presence of integrins in the differentiating cell layers (suprabasal epidermis) stimulated proliferation in the basal layer, possibly by signaling that there was a deficit in the size of the differentiated compartment, thereby requiring replenishment through increased proliferation [58]. Subsequently, the same group was able to show a correlation between the hyperproliferative marker Ki67 and the suprabasal expression of integrin in human epidermis reconstituted in culture [63]. Furthermore, these investigators delineated a possible role for suprabasally expressed integrin, through activating mitogen-activated protein kinase (MAPK), either directly or *via* IL-1α; this could be responsible for the epidermal hyperproliferation that occurs in psoriasis [64]. The advantage of this model is that candidate drugs that are integrin antagonistic can be used to test their psoriasis inhibitory effects. The limitation of this model, seen up to now, is that investigators may not be able to test the effects of many of the anti-T cell biologicals currently used in human patients with psoriasis, since the role of T cells in this model is uncertain at the present time.

Human skin-SCID mouse chimeric model

In 1996, Wrone-Smith and Nickoloff [59] reported a chimeric model of psoriasis. By grafting normal-appearing skin from human patients with psoriasis onto immunodeficient SCID mice, followed by intradermal injection of activated autologous immunocytes from the same patients, these authors were able to induce psoriasis lesions on these engrafted normal human skin samples [59]. With the impetus to delineate whether the epidermal cells or the immune cells were involved in the primary disorder in psoriasis, these authors engrafted full-thickness normal-appearing skin obtained from human psoriasis patients and healthy subjects onto SCID mice. Subsequent dermal injection of autologous peripheral blood immunocytes, isolated by density gradient centrifugation (Ficoll-Hypaque) and stimulated by IL-2 and superantigens (staphylococcal enterotoxin B and C2), induced clinical phenotype of psoriasis lesions in the engrafted non-lesional psoriatic skin [59]. Histopathological examination of the induced lesions showed characteristic features of psoriasis: hyperkeratosis, parakeratosis, loss of granular layer, acanthosis, elongation of rete ridges, Munro's microabscesses, increase of vascularity, and increase of epidermal and dermal infiltration of mononuclear cells [59]. Immunohistochemistry studies further revealed that these induced psoriatic lesions exhibited many different markers of psoriasis phenotype: T cells (CD3), T cell subsets (CD4, CD8, CD25, CD45RO), adhesion molecules [ICAM-1, VCAM-1, E-selctin (focal expression), CLA (cutaneous lymphocyte antigen)], antigen-presenting cells (CD1a, factor XIIIa), co-stimulatory molecule (CD80), macrophages (CD14, only occasionally present), and epidermal activation/proliferation markers [involucrin, β1 integrin, HLA-DR, Mac 387, keratin 16, proliferating cell nuclear antigen (PCNA)] [59]. Both epidermal and dermis of these induced skin lesions contained CD3[+] T cells, and CD4[+] and CD8[+] T cell subsets, with more CD8[+] cells in the epidermis than CD4[+] cells [59]. Full-fledged psoriasis lesions can only be induced on engrafted non-lesional psoriatic skin by autologous immune cells activated in the presence of IL-2 and superantigens, but not by unactivated autologous immune cells (except in one patient), culture conditional medium from activated autologous immune cells, combination of cytokines and superantigens, superantigens alone, or cytokine alone [59]. Subsequently, these authors showed that enriched CD4[+] T cell subset may be the most important cell type involved in the disease-induction process [65]. In addition, a T cell line that was derived from a skin lesion of a patient with psoriasis and bearing natural killer receptors CD94 and CD161 was also shown to be capable of inducing the same psoriatic skin lesions in non-lesional skin engrafted onto SCID mice [66]. Furthermore, when non-lesional psoriatic skin was engrafted onto AGR129 mice, a strain of mice deficient in type I and type II IFN receptors and the recombination-activating gene 2, spontaneous development of psoriatic skin lesions was observed. Blocking T cells led to inhibition of psoriatic skin lesion development, and supports the notion that local T cell proliferation is essential for disease induction [67]. The findings for this model therefore argue strongly that the primary disorder of psoriasis is the abnormality within the immune cells, and that the epidermal and endothelial hyperproliferations are the results of secondary activations [59]. The role of T cells is further supported by a model similar to the human skin-SCID mouse, showing that the maintenance of psoriasis lesions is dependent on T cells [68] and by the development of psoriasis after syngeneic bone marrow transplant from a psoriatic donor [69]. However, the role of epidermal abnormality cannot be totally dismissed from this model. Thorough consideration of the pathophysiology of this model indicates the dependency of disease on the epidermis from patients with psoriasis. The keratinocytes from these psoriatic patients may actually be predisposed to develop the clinical phenotype once an optimal condition is provided (in this case it could be cytokines provided by activated T cells or direct signaling from activated T cells). The advantage of this model is that it is likely the model closest to human psoriasis, due to the presence of the human skin component. The disadvantage of this model is the limited skin area available for investigational usage. The dependency of the available non-lesional skin from human patients with psoriasis and the multiple-step methods involving skin collection and engraftment, and immune cell isolation, culture, and stimulation set additional limitations on the practicality of this model.

VEGF-transgenic model

In 2003, Xia et al. [70] reported that, by transgenically delivering an important angiogenic factor VEGF to the mouse skin, an inflammatory skin condition resembling psoriasis can be generated. A transgenic construct, containing the keratin 14 expression vector and mouse VEGF cDNA, was transgenically delivered to mice with an *FVB* genetic background. The young transgenic mice developed mild skin lesions that had well-demarcated inflammatory plaques with scales, primarily located on the ear, neck, and snout [70]. When injured, the young transgenic mice developed psoriatic lesions at the injured site, resembling the Koebner's sign in human patients with psoriasis [70]. The older transgenic mice spontaneously developed skin lesions closely resembling that of human psoriasis. Histopathology examinations of these lesions showed features characteristic of human psoriasis: hyperkeratosis, parakeratosis, acanthosis, elongation of rete ridges, increase of dermal vascularity, and the presence of Munro and Kogoj microabscesses [70]. Immunohistochemistry examinations showed other features typical of psoriasis: upregulation of E-selectin, ICAM-1, VCAM-1, increase of dermal infiltration of CD4+ and CD8+ T cells, and aberrant epidermal differentiation marker K6 and K16 [70]. Interestingly, subcutaneous administration of VEGF-Trap, a potent VEGF antagonist, normalized the psoriatic phenotype induced by the VEGF transgene expression in this model, accompanied with reduction of dermal inflammatory cells, restoration of normal skin architecture, and diminution of expressions of adhesion molecules [70]. The findings of this model argued that the angiogenic factor abnormality, in association with the response of endothelial cells, may be the primary initiation factor in the development of psoriasis skin lesions. The angiogenesis process might then through its associated phenomenon of leaky blood vessels, increase migration of inflammatory cells into the skin, leading to dermal and epidermal inflammation. These inflammations, they argued, might then subsequently induce the epidermal hyperproliferative state as a secondary activity [70]. The maintenance of these psoriatic lesions appears to depend on the presence of VEGF, as these lesions were reversed by a VEGF antagonist. Thus, this model illustrated a codependence of angiogenesis and inflammation, as demonstrated by other investigations [71]. The advantage of this model is that, through it, the relationship between inflammation and angiogenesis can be delineated. Another unique benefit of this model is the possibility to investigate anti-angiogenesis candidate drugs in the treatment for psoriasis, a completely new perspective for psoriasis therapy. The limitation of this model, as currently understood, is that it may not help us delineate the role of T cells in the disease induction. Utilizing anti-T cell medication or monoclonal anti-T cell antibodies to test the reversibility of the skin lesions in this model may help eliminating the limitation of this model.

TGF-β1-transgenic model

In 2004, Li et al. [72] reported that a severe psoriasis-like skin disorder is induced by transgenically overexpressing latent wild-type TGF-β1 to the epidermal keratinocytes. The development of this model was somewhat of a surprise to some investigators, as TGF-β1 is generally considered to be an inhibitor of keratinocyte growth, rather than a promoter [72]. However, experimentally, it is now clear that TGF-β1 can exhibit either inhibitory or promoting influence on keratinocyte proliferation, depending on which level of TGF-β1 expressed [72]. For example, when constitutively active TGF-β1 was expressed in the epidermis of transgenic mice through a keratin 1 promoter, an epidermal inhibition was observed [72, 73]. On the other hand, when TGF-β1 was expressed in the epidermis of transgenic mice through a keratin 10 or keratin 6 promoter, epidermal hyperproliferation resulted [72, 74, 75]. TGF-β1 is a multifunctional cytokine that works through a heteromeric receptor complex of TGF-βRI and TGF-βRII in such a way that when TGF-β1 binds to this complex, phosphorylation of signal mediators Smad2 and Smad3 occurs, followed by the formation of a heteromeric complex between Smad4 and the phosphorylated Smad2 and Smad3, and translocation into the nucleus for the regulation of TGF-β-responsive genes [72, 76]. In the *in vitro* setting, Smad2/Smad3 activation leads to TGF-β1-induced growth inhibition of epithelial and endothelial cells [72, 77]. However, through another TGF-βRI, ALK1, which is preferentially expressed in endothelial cells, the activated ALK1 phosphorylates Smad1 and Smad5, and promotes endothelial cell proliferation and migration [72, 77]. Because TGF-β1 has such diversely observed biological activities, these investigators examined the effects of expressing it at the basal epidermis level [72]. Using a keratin 5 promoter, TGF-β1 cDNA was inserted into transgenic mice and the founders were mated with ICR or C57BL/6 background mice, which subsequently gave rise to an identical phenotype. The transgenic mice appeared normal for the first month of life, and then began to develop skin inflammatory lesions, resembling Koebner's phenomenon of human psoriasis, starting on the tagged ear at 2 months of age [72]. By 3 months of age, the transgenic mice developed focal erythematous plaques with scales, gradually involving the entire skin areas to form an "erythrodermic psoriasis" [72]. Histopathological examination revealed changes typically found in human psoriasis: acanthosis, hyperkeratosis, parakeratosis, diminished granular layers, subcorneal neutrophil microabscesses (Kogoj's), angiogenesis, and inflammatory cell infiltration [72]. Immunopathology examinations documented the increase of BM8+ macrophages, mast cells, CD4+ T cells (dermis), and CD8+ T cells (epidermis) [72]. In addition, the transgenic mice skin exhibited increased levels of mRNAs of pro-inflammatory cytokines: IL-1α, IL-1β, IL-6, IFN-γ, IL-2, and TNF-α as well as five chemokines: macrophage inflammatory protein (MIP)-1α, MIP-1β, MIP-2, IFN-γ-induced protein (IP)-10, and monocyte-chemotactic protein (MCP)-1 [72]. Furthermore, bromodeoxyuridine (BrdU) labeling experiment illustrated increased epidermal hyperproliferation in the lesional skin [72]. Substantial angiogenesis was also observed in this model. Immunostaining with anti-CD31 documented the enlargement of dermal microvasculature over a larger area covered by the vessels in the lesional skin. Immunostaining with antibody to endoglin, an accessory TGF-β receptor primarily expressed in proliferative endothelial cells [78] showed a threefold increase of endoglin in the transgenic mouse skin [72]. These findings on angiogenesis were accompanied with increased skin expressions of VEGF, and its receptors Flt-1 and Flk-1 [72, 79]. These investigators reasoned that TGF-β1, a known potent chemotactic cytokine for leukocytes, could be responsible for the initial recruitment of the inflammatory cells observed in this model. They also believed that the *in vivo* findings of epidermal hyperproliferation in the skin of these TGF-β1-transgenic mice may actually be a secondary effect. They hypothesized that the increased inflammatory cytokines and chemokines in the skin of these transgenic mice, such as IL-1, IL-6, and IL-8 (MIP-2 in the mouse system), all of which promote keratinocyte proliferation, could be derived from keratinocytes or inflammatory cells [72]. The advantage of this model is that candidate drugs that are antagonist to TGF-β-type cytokines or their receptors, can be tested for their effectiveness. The currently noted limitation of this model is the inability to delineate the role of T cells in the disease induction. Again, as mentioned above for the VEGF-transgenic model, utilizing an anti-T cell medication or monoclonal anti-T cell antibodies to test the reversibility of the skin lesions in this model may help eliminating this limitation.

Stat3-transgenic model

In 2005, Sano et al. [80] reported that a skin condition closely resembling human psoriasis can be developed by transgenically expressing a constitutively active Stat3, a transcriptional activator and signal transducer, to the skin. Stat3 is a member of a family of signal transducers and transcription activators, involved in transmitting extracellular signals to the nucleus of the cells [80–82]. Stat3 is known for its critical roles in many biological activities, including cell survival, migration, and proliferation, through regulating genes that encode cyclin D1, Bcl-xL, and some others [80–84]. Particularly, Stat3 activation is critical for skin wound healing [85]. Because Stat3 expression has been found to be upregulated in the psoriatic epidermis [80], these authors generated Stat3 transgenic mice on the *FVB* genetic background using a keratin 5 promoter to examine whether Stat3 has a role in the pathogenesis of psoriasis [80]. At birth the skin of Stat3-transgenic mice appears normal, but, by 2 weeks of age, the skin of these transgenic mice became erythematous and scaly, and they spontaneously developed scaly and hyperkeratotic lesions, starting on the tail skin and subsequently spreading to involve dorsum and hind feet [80]. The role of Stat3 in the induction and maintenance of psoriatic skin lesions was confirmed by a topically applied Stat3-specific decoy oligonucleotide, which inhibited the development of tape stripping-induced skin lesions and reversed the preexisting psoriatic lesions induced by topical treatment of TPA (12-0-tetradecanoylphorbol-13-acetate) [80]. In addition, injury of skin by full-thickness wounding induced scaly, erythematous lesions, a finding comparable to Koebner's phenomenon in human patients with psoriasis [80]. Furthermore, tape stripping of normal-appearing skin of the transgenic mice also induced well-demarcated scaly lesions. However, tape stripping alone did not result in psoriatic lesion when it was performed on normal-appearing skin of Stat3-transgenic mice engrafted onto athymic nude mice [80]. Such engrafted skin lesions can be induced by both tape stripping together with intradermal injection of activated T cells [80]. Histopathological examinations of the spontaneous or injury-induced skin lesions revealed findings characteristics of human psoriasis: acanthosis, loss of granular layer, elongation of rete ridges, increase of vascularity, hyperkeratosis, parakeratosis, and increase inflammatory cell infiltration in both epidermis and dermis [80]. Moreover, Munro's microabscesses were detected in the injury-induced lesions [80]. Examination of the immunopathology of these skin lesions further showed a predominant CD4$^+$ T cell infiltration into both dermis and epidermis, with a few CD8$^+$ T cells, primarily in the epidermis [67]. The expressions of markers of epidermal hyperproliferation characteristics of human psoriasis were also detected in these mouse psoriatic lesions, including increased Ki67, and reduced suprabasal layer keratin 1 and replaced by keratin 6 [80]. The advantage of this model is that candidate drugs that target Stat3 could be investigated for their effectiveness in this model. However, there may be a practical issue in the ultimate usefulness of Stat3 antagonist. Since Stat3 is essential for proper wound healing, targeting it may result in wound healing difficulties, especially if the psoriatic patient is an elderly individual, who naturally has lower ability to heal wounds [86, 87].

Tie2-transgenic model

Also in 2005, Voskas et al. [88] reported that a Tie2-transgenic mouse line, which can exhibit skin condition resembling human psoriasis. Using a binary transgenic approach, in which expression of Tie2 can be conditionally regulated by the presence (or absence) of tetracycline analogs, these investigators generated a psoriasis-like skin condition that initiates at 5 days of age and persists throughout adulthood [52, 88]. Although the effects of Tie2 on the microvasculature are not yet clear, it is now known that Tie2 is required for vessel maintenance, and it is thought that in concert with VEGF it regulates the expression of angiopoietin-2, thereby facilitating vascular destabilization and blood vessel sprouting by antagonizing the stabilizing function of angiopoietin-1 [88–93]. Importantly, expression of Tie2 and the angiopoietins has been documented to be upregulated in human psoriatic lesions [94]. In this model, conditional overexpression of Tie2 by a promoter resulted in a reversible skin disorder closely resembling human psoriasis. By 5 days after birth, the double transgenic mice were noted to have extensive erythema and loosely adherent silvery scales. In the adult transgenic mice, thicken erythematous skin lesions developed on the ears, snout, peri-orbital skin, knuckles of the legs, and nape of the neck [88]. In addition, these transgenic mice tended to develop lesions at trau-matized sites (Koebner's sign), and have yellow-brown discoloration of the nails, both of these are characteristic findings in human psoriasis [88]. Interestingly, these investigators showed not only the presence of Koebner's sign, but also Auspitz sign, a highly specific psoriatic phenomenon in which, in these transgenic mice, the pres-ence of multiple bleeding point is observed when a scale is physically removed [88]. Histopathological examination of the skin lesions of the transgenic mice demon-strated typical features of human psoriasis: acanthosis, hyperkeratosis, parakerato-sis, elongation of rete ridges, increase dermal vasculature, dermal infiltration of mononuclear cells, and neutrophil microabscesses within the epidermis [88]. Immunopathology involved increases of dermal and epidermal infiltration of CD3[+] T cells into the skin lesions of these transgenic mice [88]. Cyclosporine A, an anti-T cell medication, was able to substantially improve the disease phenotype, along with reversal of the histological changes in epidermal thickness, vasculature, and inflammatory cell infiltration [88]. However, it is not clear whether the cyclosporine-induced resolution of the skin lesion is due to its effect on T cells, endothelial cells, epithelial cells, or a combination of these, since cyclosporine A could affect any one of these three cell types [88, 95, 96]. One advantage of this model is the unique ability to regulate Tie2 expression levels by conditional expres-sion so that, by simultaneous monitoring the disease progression, the investigators could utilize this model to dissect the pathomechanism of the disease, and therefore delineate potential targets for intervention. The other advantages and limitations are similar to the VEGF-transgenic mice discussed above.

Therapeutic tests on animal models

Several of the animal models included in this chapter have been used for testing potential drug or gene therapy candidates. Table 4 exhibits the results of these tests and provides a helpful guide to the potential of these mouse models of atopic der-matitis and psoriasis as tools for candidate medication/treatment screening. Since the majority of these models were developed recently, the absence of published reports on therapy performed on these models does not imply the absence of potentials.

Table 4 - Therapeutic potentials of mouse models of atopic dermatitis and psoriasis

Disease	Model	Treatment performed	Results	Refs.
Atopic dermatitis	NC/Nga	Psoralens + UVA[a]	Improved	[97]
	NC/Nga	Konjac-glucomannan	Improved	[98]
	NC/Nga	Immunomodulator FTY 720	Prevented	[99]
	NC/Nga	IL-10 antisense oligo[b]	Improved	[100]
	NC/Nga	Topical tacrolimus	Inhibited	[101]
	NC/Nga	Herbal medicine Hochu-ekki-to	Suppressed	[102]
	NC/Nga	Anti-IL-18 Abs[c]	Failed	[103]
	NC/Nga	NF-κB decoy oligo	Prevented/ improved	[104]
	NC/Nga	Chymase inhibitor	Improved	[105]
	NC/Nga	TGF-β1	Suppressed	[106]
	NC/Nga	Persimmon leaf extract	Improved	[107]
Psoriasis	SCID-Hu[d]	Anti-CD11a mAbs	Improved	[108]
	SCID-Hu	Cyclosporin A	Improved	[108]
	SCID-Hu	Clobetasol propionate	Improved	[108]
	SCID-Hu	Anti-IL-15 Abs	Resolved	[109]
	VEGF-Tg	VEGF Trap	Reversed	[70]
	K5-Stat3	Decoy oligo	Improved	[110]

[a] UVA, ultraviolet light A
[b] oligo, oligonucleotides
[c] Abs, antibodies
[d] SCID-Hu, severe combined immunodeficiency mice-human skin graft

Summary

In this chapter, animal models of two common inflammatory skin disorders, atopic dermatitis and psoriasis, have been described, and their potential utility for understanding pathophysiology of the diseases and for pre-clinical investigation of candidate drug efficacy are discussed. The basic rationales for selecting these two diseases are that these diseases represent chronic inflammatory skin conditions impacting our society in the most significant way, that the *in vivo* models of inflammatory skin diseases are most useful to investigate complex biological interactions in diseases of a chronic nature (rather than disease of an acute nature), and that the included models for these diseases have been developed recently (with the earliest atopic dermatitis and psoriasis models developed in 1997 and 1995, respectively) and therefore containing the most updated information. Moreover, the models presented in this chapter, in comparison to prior models, are much closer to the human diseases on clinical, histological, and immunological grounds, and are therefore better models for the purpose of studying the disease mechanisms and of screening new therapeutic candidates. From the available information on the animal models of atopic dermatitis thus far, it is safe to conclude that atopic dermatitis is an allergic skin disease that results from overresponding to environmental antigens, involving primarily the Th2 arm of the immune system for the initiation of the disease, but also involving the Th1 arm as well. The common histopathological and immunopathological findings in all atopic dermatitis models described in this chapter include acanthosis, spongiosis, and increased dermal infiltration of mast cells, eosinophils, and CD4+ T cells (Tab. 2). Most of the currently available animal models of atopic dermatitis can be considered as "IgE-associated" or "extrinsic" models, which are marked by an increase of either total serum IgE or antigen-specific IgE, while one of the models could become a "non-IgE-associated" or "intrinsic" model, pending full characterization of the model. Since each model system has its advantages, as well as its limitations, the readers are advised to select the model system that suits their primary investigation goal the best. As for the models of psoriasis, a common pathophysiological pathway seems to emerge from the available data concerning these models with a seeming dichotomy of mechanisms. It seems possible that the psoriatic skin is predisposed to over-reactive response to physical injury. When skin is injured, the naturally occurring inflammatory response to the injury mobilizes the immune system, involving various cytokines, chemokines, inflammatory cells, and the complement system, for the completion of wound healing. Unlike the wound healing process occurring in normal individuals, with its complex and tightly regulated mechanisms that ensure that proper wound healing does not result in aberrant epidermal and endothelial growth, the psoriatic patients responded to injury by increases in both epidermal and endothelial hyperproliferation; the later might also lead to an increase of inflammatory cell infiltration. This exaggerated wound healing tendency in the patients with psoriasis might be so great that an ordinary immune response to a common pathogen challenge, such as streptococcal infection, would trigger the epidermal and endothelial hyperproliferation. Histopathological and immunopathological features observed in all of the models discussed in this chapter include acanthosis, hyperkeratosis, parakeratosis, increased vascularity, and increased dermal CD4+ T cells. In most models there is also a loss of granular layer and they are marked by the presence of Munro's and/or Kogoj's microabscesses of neutrophils (Tab. 3). Like the atopic dermatitis models, each model system of psoriasis has its advantages, as well as its limitations, and naturally readers are advised to select the model system that is most suitable for their chief investigation aim.

Acknowledgements

This work is supported in part by a grant from National Institutes of Health, USA (R01 AR47667, L. S. Chan) and Veterans Affairs Medical Center, Chicago, IL, USA (L. S. Chan).

References

1   Freedberg IM, Eisen AZ, Wolff K, Austen KF, Goldsmith LA, Katz SI (eds) (2003) *Fitzpatrick's dermatology in general medicine*, 6th edn. McGraw-Hill, New York

2   The Lewin Group (2005) *The burden of skin diseases*. Society for Investigative Dermatology & American Academy of Dermatology Association, Cleveland

3   Chan LS (2004) Comparative structure and function of the skin: overview of structures and components. In: LS Chan (ed): *Animal Models of Human Inflammatory Skin Diseases*. CRC Press, Boca Raton, 3–17

4   Chan LS, Gordon KB (2004) Mouse immune system. In: LS Chan (ed): *Animal Models of Human Inflammatory Skin Diseases*. CRC Press, Boca Raton, 119–140

5   Hultsch T, Kapp A, Spergel J (2005) Immunomodulation and safety of topical calcineurin inhibitors for the treatment of atopic dermatitis. *Dermatology* 211: 174–187

6   Peterson J, Chan LS (2006) Comprehensive management guidelines for atopic dermatitis. *Dermatol Nursing, in press*

7   Lebwohl M (2005) A clinician's paradigm in the treatment of psoriasis. *J Am Acad Dermatol* 53: S59–69

8   Van de Kerkhof PC, Kragballe K (2005) Recommendations for the topical treatment of psoriasis. *J Eur Acad Dermatol* 19: 495–499

9   Dando TM, Wellington K (2005) Topical tazarotene: a review of its use in the treatment of plaque psoriasis. *Am J Clin Dermatol* 6: 255–272

10 Lee E, Koo J, Berger T (2005) UVB phototherapy and skin cancer risk: a review of the literature. *Int J Dermatol* 44: 355–360

11 Bandow GD, Koo JY (2004) Narrow-band ultraviolet B radiation: a review of the current literature. *Int J Dermatol* 43: 555–561

12 Van Zander J, Orlow SJ (2005) Efficacy and safety of oral retinoids in psoriasis. *Expert Opin Drug Saf* 4: 129–138

13 Warren RB, Griffiths CE (2005) The potential of pharmacogenetics in optimizing the use of methotrexate for psoriasis. *Br J Dermatol* 153: 869–873

14 Behnam SM, Behnam SE, Koo JY (2005) Review of cyclosporine immunosuppressive safety data in dermatology patients after two decades of use. *J Drugs Dermatol* 4: 189–194

15 Sterry W, Barker J, Boehncke WH, Bos JD, Chimenti S, Christophers E, De La Brassinne M, Ferrandiz C, Griffiths C, Katsambas A et al (2004) Biological therapies in the sys-

temic management of psoriasis: International Consensus Conference. *Br J Dermatol* 151 Suppl 69: 3–17

16 Weinberg JM, Bottino CJ, Lindholm J, Buchholz R (2005) Biologic therapy for psoriasis: an update on the tumor necrosis factor inhibitors infliximab, etanercept, and adalimumab, and the T-cell-targeted therapies efalizumab and alefacept. *J Drugs Dermatol* 4: 544–555

17 Thielen AM, Kuenzli S, Saurat JH (2005) Cutaneous adverse events of biological therapy for psoriasis: review of the literature. *Dermatology* 211: 209–217

18 Leung DYM, Bieber T (2003) Atopic dermatitis. *Lancet* 361: 151–1160

19 Leung DYM, Boguniewicz M, Howell MD, Nomura I, Hamid QA (2004) New insights into atopic dermatitis. *J Clin Invest* 113: 651–657

20 Hanifin JM, Rajka G (1980) Diagnostic features of atopic dermatitis. *Acta Dermatolvener (Stockholm)* (Suppl) 92: 44–47

21 Wuthrich B, Schmid-Grendelmeier P (2003) The atopic eczema/dermatitis syndrome. Epidemiology, natural course, and immunology of the IgE-associated ("extrinsic") and the nonallergic ("intrinsic") AEDS. *J Investig Allergol Clin Immunol* 13: 1–5

22 Flohr C, Jonansson SG, Wahlgren CF, Willimans H (2004) How atopic is atopic dermatitis? *J Allergy Clin Immunol* 114: 150–158

23 Novak N, Bieber T, Leung DYM (2003) Immune mechanisms leading to atopic dermatitis. *J Allergy Clin Immunol* 112: S128–39

24 Gutermuth J, Ollert M, Ring J, Behrendt H, Jakob T (2004) Mouse models of atopic eczema critically evaluated. *Int Arch Allergy Immunol* 135: 262–276

25 Matsuda H, Watanabe N, Geba GP, Sperl J, Tsudzuki M, Hiroi J, Matsumoto M, Ushio H, Saito S, Askenase PW, Ra C (1997) Development of atopic dermatitis-like skin lesion with IgE hyperproduction in NC/Nga mice. *Int Immunol* 9: 461–466

26 Jones HE, Inoue JC, McGertly JL, Lewis CW (1975) Atopic dermatitis and serum immunoglobulin-E. *Br J Dermatol* 92: 17–25

27 Mihm MC, Jr, Soter NA, Dvorak HF, Austen KF (1976) The structure of normal skin and the morphology of atopic eczema. *J Invest Dermatol* 67: 305–312

28 Hamid Q, Boguniewicz M, Leung DY (1994) Differential in situ cytokine gene expression in acute versus chronic atopic dermatitis. *J Clin Invest* 94: 870–876

29 Hamid Q, Naseer T, Minshall EM, Song YL, Boguniewicz M, Leung DY (1996) *In vivo* expression of IL-12 and IL-13 in atopic dermatitis. *J Allergy Clin Immunol* 98: 225–231

30 Grewe M, Gyufko K, Schopf E, Krutmann J (1994) Lesional expression of interferon-gamma in atopic dermatitis. *Lancet* 343: 25–26

31 Vestergaard C, Yoneyama H, Murai M, Nakamura K, Tamaki K, Terashima Y, Imai T, Yoshie O, Irimura T, Mizutani H, Matsushima K (1999) Overproduction of Th2-specific chemokines in NC/Nga mice exhibiting atopic dermatitis-like lesions. *J Clin Invest* 104: 1097–1105

32 Matsumoto M, Ra C, Kawamoto K, Sato H, Itakura A, Sawada J, Ushio H, Suto H, Mitsuishi K, Hikasa Y, Matsuda H (1999) IgE hyperproduction through enhanced tyro-

sine phosphorylation of Janus Kinase 3 in NC/Nga mice, a model for human atopic dermatitis. *J Immunol* 162: 1056–1063

33 Aioi A, Tonogalto H, Suto H, Hamada K, Ra CR, Ogawa H, Maibach H, Matsuda H (2001) Impairment of skin barrier function in NC/Nga Tnd mice as possible model for atopic dermatitis. *Br J Dermatol* 144: 12–18

34 Iwasaki T, Tanaka A, Itakura A, Yamashita N, Ohta K, Matsuda H, Onuma M (2001) Atopic NC/Nga mice as a model for allergic asthma: severe allergic responses by single intranasal challenge with protein antigen. *J Vet Med Sci* 63: 413–419

35 Sasakawa T, Higashi Y, Sakuma S, Hirayama Y, Sasakawa Y, Ohkubo Y, Goto T, Matsumoto M, Matsuda H (2001) Atopic dermatitis-like skin lesions induced by topical application of mite antigens in NC/Nga mice. *Int Arch Allergy Immunol* 126: 239–247

36 Kawamoto K, Matsuda H (2004) Spontaneous mouse model of atopic dermatitis in NC/Nga mice. In: LS Chan (ed.): *Animal Models of Human Inflammatory Skin Diseases*. CRC Press, Boca Raton, 371–386

37 Wang LF, Lin JY, Hsich KH, Lin RH (1996) Epicutaneous exposure of protein antigen induces a predominant Th2-like response with high IgE production in mice. *J Immunol* 156: 4077–4082

38 Spergel JM, Mizoguchi E, Brewer JP, Martin TR, Bhan AK, Geha RS (1998) Epicutaneous sensitization with protein antigen induces localized allergic dermatitis and hyper-responsiveness to methacholine after single exposure to aerosolized antigen in mice. *J Clin Invest* 101: 1614–1622

39 Spergel JM, Mizoguchi E, Oettgen H, Bhan AK, Geha RS (1999) Roles of TH1 and TH2 cytokines in a murine model of allergic dermatitis. *J Clin Invest* 103: 1103–1111

40 Spergel JM (2004) Experimental mouse model of atopic dermatitis: induction by epicutaneous application of allergen. In: LS Chan (ed): *Animal Models of Human Inflammatory Skin Diseases*. CRC Press, Boca Raton, 417–426

41 Woodward AL, Spergel JM, Alenius H, Mizoguchi E, Bhan AK, Castigli E, Brodeur SR, Oettgen HC, Geha RS (2001) An obligate role for T-cell receptor alphabeta+ T cells but not T-cell receptor gammadelta+ T cells, B cells, or CD40/CD40L interactions in a mouse model of atopic dermatitis. *J Allergy Clin Immunol* 107: 359–366

42 Li XM, Kleiner G, Huang CK, Lee SY, Schofield B, Soter NA, Sampson HA (2001) Murine model of atopic dermatitis associated with food hypersensitivity. *J Allergy Clin Immunol* 107: 693–702

43 Li XM, Sampson HA (2004) Experimental mouse model of atopic dermatitis: induction by oral allergen. In: LS Chan (ed): *Animal Models of Human Inflammatory Skin Diseases*. CRC Press, Boca Raton, 389–415

44 Chan LS, Robinson N, Xu L (2001) Expression of interleukin-4 in the epidermis of transgenic mice results in a pruritic inflammatory skin disease: An experimental animal model to study atopic dermatitis. *J Invest Dermatol* 117: 977–983

45 Chen L, Martinez O, Overbergh L, Mathieu C, Prabhakar BS, Chan LS (2004) Early up-regulation of Th2 cytokines and late surge of Th1 cytokines in an atopic dermatitis model. *Clin Exp Immunol* 138: 375–383

46 Chen L, Martinez O, Venkataramani P, Lin SX, Prabhakar BS, Chan LS (2005) Correlation of disease evolution with progressive inflammatory cell activation and migration in the IL-4 transgenic mouse model of atopic dermatitis. *Clin Exp Immunol* 139: 189–201

47 Chen L, Lin SX, Overbergh L, Mathieu C, Chan LS (2005) The disease progression in the keratin 14 IL-4-transgenic mouse model of atopic dermatitis parallels the upregulations of B cell activation molecules, proliferation, and surface and serum IgE. *Clin Exp Immunol* 142: 21–30

48 Chan LS (2004) Experimental mouse model of atopic dermatitis by transgenic induction. In: LS Chan (ed): *Animal Models of Human Inflammatory Skin Diseases*. CRC Press, Boca Raton, 387–398

49 Agha-Majzoub R, Becker RP, Schraufnagel DE, Chan LS (2005) Angiogenesis, the major abnormality of the Keratin 14 IL-4-transgenic mouse model of atopic dermatitis. *Microcirculation* 12: 455–476

50 Lee SY, Paik SY, Chung SM (2005) Neovastat (AE-941) inhibits the airway inflammation and hyperresponsiveness in a murine model of asthma. *J Microbiol* 43: 11–16

51 Watanabe H, Mamelak AJ, Wang B, Howell BG, Freed I, Esche C, Nakayama M, Nagasaki G, Hicklin DJ, Kerbel RS, Sauder DN (2004) Anti-vascular endothelial growth factor receptor-2 (Flk-1/KDR) antibody suppresses contact hypersensitivity. *Exp Dermatol* 13: 671–681

52 Chan LS (2004) Molecular biological manipulation of the immune system by transgenic techniques. In: LS Chan (ed): *Animal Models of Human Inflammatory Skin Diseases*. CRC Press, Boca Raton, 187–195

53 Rzany B, Naldi L, Schafer T, Stern R, Williams H (1998) The diagnosis of psoriasis: diagnostic criteria. *Br J Dermatol* 138: 917

54 Christophers E, Mrowietz U (2003) Psoriasis. In: Freedberg IM, Eisen AZ, Wolff K, Austen KF, Goldsmith LA, Katz SI (eds): *Fitzpatrick's dermatology in general medicine*, 6th edn. McGraw-Hill, New York, 407–427

55 Stern RS, Wu J (1996) Psoriasis. In: KA Arndt, PE LeBoit, JK Robinson, BU Wintroub (eds): *Cutaeneous medicine and surgery: An integrated program in dermatology*. Saunders, Philadelphia, 295–321

56 Leigh IM, Navsaria H, Purkis PE, McKay IA, Bowden PE, Riddle PN (1995) Keratins (K16 and K17) as markers of keratinocyte hyperproliferation in psoriasis *in vivo* and *in vitro*. *Br J Dermatol* 133: 501–511.

57 Hertle MD, Kubler MD, Leigh IM, Watt FM (1992) Aberrant integrin expression during epidermal wound healing and in psoriastic epidermis. *J Clin Invest* 89: 1892–1901

58 Caroll JM, Romero MR, Watt FM (1995) Suprabasal integrin expression in the epidemis of transgenic mice results in development defects and a phenotype resembling psoriasis. *Cell* 83: 957–968

59 Wrone-Smith T, Nickoloff BJ (1996) Dermal injection of immunocytes induces psoriasis. *J Clin Invest* 98: 1878–1887

60   Watt FM, Hertle MD (1994) Keratinocyte integrins. In: IM Leigh, EB Lane, FM Watt (eds): *The keratinocyte handbook*. Cambridge University Press, Cambridge, 153–164

61   Hotchin NA, Kovach NL, Watt FM (1993) Functional downregulation of α5β1 integrin in keratinocytes is reversible but commitment to terminal differentiation is not. *J Cell Sci* 106: 1131–1138

62   Hotchin NA, Gandarilias A, Watt FM (1995) Regulation of cell surface β1 integrin levels during keratincocyte terminal differentiation. *J Cell Biol* 128: 1209–1219

63   Rikimaru K, Moles JP, Watt FM (1997) Correlation between hyperproliferation and suprabasal integrin expression in human epidermis reconstituted in culture. *Exp Dermatol* 6: 214–221

64   Haase I, Hobbs RM, Romero MR, Broad S, Watt FM (2001) A role for mitogen-activated protein kinase activation by integrins in the pathogenesis of psoriasis. *J Clin Invest* 108(4): 527–536

65   Nickoloff BJ, Wrone-Smith T (1999) Injection of pre-psoriatic skin with CD4+ T cells induces psoriasis. *Am J Pathol* 155: 145–158

66   Nickoloff BJ, Bonish B, Huang BB, Porcelli SA (2000) Characterization of a T cell line bearing natural killer receptors and capable of creating psoriasis in a SCID mouse model system. *J Dermatol Sci* 24(3): 212–225

67   Boyman O, Hefti HP, Conrad C, Nickoloff BJ, Suter M, Nestle FO (2004) Spontaneous development of psoriasis in a new animal model shows an essential role for resident T cells and tumor necrosis factor-alpha. *J Exp Med* 199(5): 731–736

68   Gilhar A, David M, Ullmann Y, Berkutski T, Kalish RS (1997) T-lymphocyte dependence of psoriatic pathology in human psoriatic skin grafted to SCID mice. *J Invest Dermatol* 109(3): 283–288

69   Snowden JA, Heaton DC (1997) Development of psoriasis after syngeneic bone marrow transplant from psoriatic donor: further evidence for adoptive autoimmunity. *Br J Dermatol* 137(1): 130–132

70   Xia YP, Li B, Hylton D, Detmar M, Yancopoulos GD, Rudge JS (2003) Transgenic delivery of VEGF to mouse skin leads to an inflammatory condition resembling human psoriasis. *Blood* 102: 161–168

71   Jackson JR, Seed MP, Kircher CH, Willoughby DA, Winkler JD (1997) The codependence of angiogenesis and chronic inflammation. *FASEB J* 11: 457–465

72   Li AG, Wang D, Feng XH, Wang XJ (2004) Latent TGFβ1 overexpression in keratinocytes results in a severe psoriasis-like skin disorder. *EMBO J* 23: 1770–1781

73   Sellheyer K, Bickenbach JR, Rothnagel JA, Bundman D, Longley MA, Krieg T, Roche NS, Roberts AB, Roop DR (1993) Inhibition of skin development by overexpression of transforming growth factor beta 1 in the epidermis of transgenic mice. *Proc Natl Acad Sci USA* 90: 5237–5241

74   Cui W, Fowlis DJ, Cousins FM, Duffie E, Bryson S, Balmain A, Akhurst RJ (1995) Concerted action of TGF-beta 1 and its type II receptor in control of epidermal homeostasis in transgenic mice. *Genes Dev* 9: 945–955

75   Fowlis DJ, Cui W, Johnson SA, Balmain A, Akhurst RJ (1996) Altered epidermal cell

growth control *in vivo* by inducible expression of transforming growth factor beta 1 in the skin of transgenic mice. *Cell Growth Differ* 7: 679–687

76 Miyazono K, Kusanagi K, Inoue H (2001) Divergence convergence of TGF-beta/BMP signaling. *J Cell Physiol* 187: 265–276

77 Goumans MJ, Valdmarsdottir G, Itoch S, Rosendahl A, Sideras P, Ten Dijke P (2002) Balancing the activation state of the endothelium via two distinct TGF-beta type I receptors. *EMBO J* 21: 1743–1753

78 Rulo HF, Westphal JR, van de Kerkhof PC, de Waal RM, van Vlijmen IM, Ruiter DJ (1995) Expression of endoglin in psoriatic involved and uninvolved skin. *J Dermatol Sci* 10: 103–109

79 Shibuya M (2003) Vascular endothelial growth factor receptor-2: Its unique signaling and specific ligand, VEGF-E. *Cancer Sci* 94: 751–756

80 Sano S, Chan KS, Carbajal S, Clifford J, Peavey M, Kiguchi K Itami S, Nickoloff BJ, DiGiovanni J (2005) Stat3 links activated keratinocytes and immunocytes required for development of psoriasis in a novel transgenic mouse model. *Nat Med* 11: 43–49

81 Leonard WJ, O'Shea JJ (1998) Jaks and STATs: biological implications. *Annu Rev Immunol* 16: 293–322

82 Levy ED, Darnell JE, Jr (2002) Stats: transcriptional control and biological impact. *Nat Rev Cell Biol* 3: 651–662

83 Hirano T, Ishihara K, Hibi M (2000) Roles of STAT3 in mediating the cell growth, differentiation and survival signals relayed through the IL-6 family of cytokine receptors. *Oncogene* 19: 2548–2556

84 Turkson J, Jove R (2000) STAT proteins: novel molecule targets for cancer drug discovery. *Oncogene* 19: 6613–6626

85 Sano S, Itami S, Takeda K, Tarutani M, Yamaguchi Y, Miura H, Yoshikawa K, Akira S, Takeda J (1999) Keratinocyte-specific ablation of Stat3 exhibits impaired skin remodeling, but does not affect skin morphogenesis. *EMBO J* 18: 4657–4668

86 Reed MJ, Ferara NS, Vernon RB (2001) Impaired migration, integrin function, and actin cytoskeletal organization in dermal fibroblasts from a subset of aged human donors. *Mech Ageing Dev* 122: 1203–1220

87 Gosain A, DiPietro LA (2004) Aging and wound healing. *World J Surg* 28: 321–326

88 Voskas D, Jones N, Van Slyke P, Struk C, Chang W, Haninec A, Babichev YO, Tran J, Master Z et al (2005) A cyclosporine-sensitive psoriasis-like disease produced in Tie2 transgenic mice. *Am J Pathol* 166: 843–855

89 Sato TN, Tozawa Y, Deutsch U, Wolburg-Buchholz K, Fujiwara Y, Gendron-Maguire M, Gridley T, Wolburg H, Risau W, Qin Y (1995) Distinct roles of the receptor tyrosine kinases Tie-1 and Tie-2 in blood vessel formation. *Nature* 376: 70–74

90 Maisonpierre PC, Suri C, Jones PF, Bartunkova S, Wiegand SJ, Radziejewski C, Compton D, McClain J, Aldrich TH, Papadopoulos N et al (1997) Angiopoietin-2, a natural antagonist for Tie2 that disrupts *in vivo* angiogenesis. *Science* 277: 55–60

91 Ward N, Dumont DJ (2002) The angiopoietins and Tie2/Tek: adding to the complexity of cardiovascular development. *Semin Cell Dev Biol* 1: 19–27

92 Puri MC, Partanen J, Rossant J, Bernstein A (1999) Interaction of the TEK and TIE receptor tyrosine kinases during cardiovascular development. *Development* 126: 4569–4580

93 Jones N, Voskas D, Master Z, Sarao R, Jones J, Dumont DJ (2001) Rescue of the early vascular defects in Tek/Tie2 null mice reveals an essential survival function. *EMBO Rep* 2: 438–445

94 Kuroda K, Sapadin A, Shoji T, Fleischmajer R, Lebwohl M (2001) Altered expression of angiopoietins and Tie2 endothelium receptor in psoriasis. *J Invest Dermatol* 116: 713–720

95 Hernandez GL, Volpert OV, Inguez MA, Lorenzo E, Martinez-Martinez S, Grau R, Fresno M, Redondo JM (2001) Selective inhibition of vascular endothelial growth factor-mediated angiogenesis by cyclosporin A: roles of nuclear factor of activated T cells and cycloxygenase 2. *J Exp Med* 193: 607–620

96 Nickoloff BJ, Fisher GJ, Mitra RS, Voorhees JJ (1988) Additive and synergistic antiproliferative effects of cyclosporine A and gamma interferon on cultured human keratinocytes. *Am J Pathol* 131: 12–18

97 Miyauchi-Hashimoto H, Okamoto H, Sugihara A, Horio T (2005) Therapeutic and prophylactic effects of PUVA photochemotherapy on atopic dermatitis-like lesions in NC/Nga mice. *Photodermatol Photoimmunol Photomed* 21: 125–130

98 Onishi N, Kawamoto S, Nishimura M, Nakano T, Aki T, Shigeta S, Shimizu H, Hashimoto K, Ono K (2004) The ability of konjac-glucomannan to suppress spontaneously occurring dermatitis in NC/Nga mice depends upon the particle size. *Biofactors* 21: 163–166

99 Kohno T, Tsuji T, Hirayama K, Watabe K, Matsumoto A, Kohno T, Fujita T (2004) A novel immunomodulator, FTY720, prevents spontaneous dermatitis in NC/Nga mice. *Biol Pharm Bull* 27: 1392–1396

100 Sakamoto T, Miyazaki E, Aramaki Y, Arima H, Takahashi M, Kato Y, Koga M, Tsuchiya S (2004) Improvement of dermatitis by iontophoretically delivered antisense oligonucleotides for interleukin-10 in NC/Nga mice. *Gene Ther* 11: 317–324

101 Sasakawa T, Higashi Y, Sakuma S, Hirayama Y, Sasakawa Y, Ohkubo Y, Mutoh S (2004) Topical application of FK506 (tacrolimus) ointment inhibits mite antigen-induced dermatitis by local action in NC/Nga mice. *Int Arch Allergy Immunol* 133: 55–63

102 Kobayashi H, Mizuno N, Kutsuna H, Teramae H, Ueoku S, Onoyama J, Yamanaka K, Fujita N, Ishii M (2003) Hochu-ekki-to suppresses development of dermatitis and elevation of serum IgE level in NC/Nga mice. *Drugs Exp Clin Res* 29: 81–84

103 Higa S, Kotani M, Matsumoto M, Fujita A, Hirano T, Suemura M, Kawase I, Tanaka T (2003) Administration of anti-interleukin 18 antibody fails to inhibit development of dermatitis in atopic dermatitis-model mice NC/Nga. *Br J Dermatol* 149: 39–45

104 Nakamura H, Aoki M, Tamai K, Oishi M, Ogihara T, Kaneda Y, Morishita R (2002) Prevention and regression of atopic dermatitis by ointment containing NF-κB decoy oligodeoxynucleotides in NC/Nga atopic mouse model. *Gene Ther* 9: 1221–1229

105 Watanabe N, Tomimori Y, Saito K, Miura K, Wada A, Tsudzuki M, Fukuda Y (2002) Chymase inhibitor improves dermatitis in NC/Nga mice. *Int Arch Allergy Immunol* 128: 229–234

106 Sumiyoshi K, Nakao A, Ushio H, Mitsuishi K, Okumura K, Tsuboi R, Ra C, Ogawa H (2002) Transforming growth factor-beta1 suppresses atopic dermatitis-like skin lesions in NC/Nga mice. *Clin Exp Allergy* 32: 309–314

107 Matsumoto M, Kotani M, Fujita A, Higa S, Kishimoto T, Suemura M, Tanaka T (2002) Oral administration of persimmon leaf extract ameliorates skin symptoms and transepidermal water loss in atopic dermatitis model mice, NC/Nga. *Br J Dermatol* 146: 221–227

108 Zeigler M, Chi Y, Tumas DB, Bodary B, Tang H, Varani J (2001) Anti-CD11a ameliorates disease in the human psoriatic skin-SCID mouse transplant model: Comparison of antibody to CD11a with Cyclosporin A and clobetasol propionate. *Lab Invest* 81: 1253–1261

109 Villadsen LS, Schuarman J, Beurskens F, Dan TN, Dagnaes-Hansen F, Skov L, Rygaard J, Voorhorst-Ogink MM, Gerristen AF, van Dijk MA et al (2003) Resolution of psoriasis upon blockade of IL-15 biological activity in a xenograft mouse model. *J Clin Invest* 112: 1571–1580

110 Sano S, Chan KS, Carbajal S, Clfford J, Peavey M, Kiguchi K, Itami S, Nickoloff BJ, DiGiovanni J (2005) Stat3 links activated keratinocytes and immunocytes required for development of psoriasis in a novel transgenic mouse model. *Nat Med* 11: 43–49

In vivo models of neurogenic inflammation

Pierangelo Geppetti[1,2], Serena Materazzi[1], Paola Nicoletti[1] and Marcello Trevisani[2]

[1]Clinical Pharmacology Unit, Department of Critical Care Medicine and Surgery, University of Florence, Florence, Italy; [2]Center of Excellence for the Study of Inflammation, University of Ferrara, Ferrara, Italy

Introduction

Principles of neurogenic inflammation

Peripheral tissue response to injury is, in part, mediated by neurogenic inflammation. This specific type of the inflammatory response is attributed to a local effector function of primary afferent terminals [1–3]. The sensory neurons involved in neurogenic inflammation have small and dark cell bodies, located in dorsal root, trigeminal and vagal ganglia and non-myelinated or thinly myelinated fibers of the C and A-δ type, respectively [4]. Nerve fibers of primary sensory neurons terminate centrally within the lamina I and II of the dorsal spinal cord and brainstem, whereas peripheral terminals are almost ubiquitously distributed to all tissues and organs with a marked prevalence around blood vessels. The mature expression of the phenotype of neurons that mediate neurogenic inflammation, which is dependent on nerve growth factor [5], includes the expression of a non-selective cation channel of the transient receptor potential (TRP) family of channels that, being sensitive to vanilloid molecules, has been termed vanilloid 1 (TRPV1) [6, 7]. Capsaicin, the pungent principle contained in the plants of the genus *Capsicum* has the ability to selectively stimulate the TRPV1, thus causing neuronal excitation, neuropeptide release and neurogenic inflammatory responses [8–10]. The ability of capsaicin to excite sensory neurons is accompanied by another useful pharmacological feature of this compound. If neurons or their terminal fibers are exposed to elevated doses of capsaicin *in vivo*, or high concentrations *in vitro*, maximal TRPV1 stimulation results in an exaggerated influx of ions into the nerve terminal/neuronal cells, causing a series of phenomena that span from desensitization to neuronal cell death [8, 9]. Thus, the unique property to produce selective stimulation/defunctionalization of the subpopulation of primary sensory neurons that mediate neurogenic inflammation has made capsaicin an invaluable tool for investigating the functions exerted by these neurons and to uncover the existence of the inflammatory response mediated by this neurogenic mechanism.

In Vivo Models of Inflammation, Vol. II, edited by Christopher S. Stevenson, Lisa A. Marshall
and Douglas W. Morgan

The ability of TRPV1-expressing sensory neurons to mediate neurogenic inflammation relies on the fact that these neurons express and release a large variety of neuropeptides. However, Ca^{2+}-dependent release and the associated biological functions have been determined only for the calcitonin gene-related peptide (CGRP) [11] and for the tachykinins, substance P (SP) and neurokinin A (NKA) [12]. Apart from a few exceptions, neurogenic inflammation encompasses the biological responses produced by stimulation by CGRP, SP or NKA of their specific receptors on effector cells. SP and NKA activate three distinct tachykinin receptor subtypes, the NK1, NK2 and NK3 receptors [13]. These receptors produce cell excitation by coupling to the a $G_{q/11}$ protein and thus, increase IP3 and Ca^{2+} levels. NK1 receptors are particularly abundant at the vascular level, where, in the endothelium, they mediate nitric oxide (NO)-dependent arterial vasodilatation, and in the venules an increased leakage of plasma proteins. NK2 receptors have a plentiful expression on smooth muscle cells (in the airways, urinary tract, etc.) where they mediate contractile responses. Finally, NK3 receptors are less diffused and present in peripheral ganglia, veins and in few other tissues [13]. The receptors for CGRP/amylin that have been cloned and characterized so far consist of a seven-transmembrane G protein-coupled calcitonin-like (CL) receptor in association with one of three single membrane-spanning receptor activity-modifying protein (RAMP)s. RAMP1 associated with CL produces a CGRP receptor (CGRP1) that is antagonized by the CGRP antagonist $CGRP_{8-37}$ [14]. All CGR/amylin receptors increase intracellular levels of cyclic AMP.

Neurogenic inflammation is particularly prominent at the vascular level where its main features are vasodilatation of arterioles, plasma protein extravasation in post-capillary venules, and leukocyte adhesion to endothelial cells of venules. Non-vascular effects of neurogenic inflammation include a large variety of diverse biological actions that differ according to the tissue and the mammal species under investigation. In the airways, prominent extravascular actions mediated by neurogenic mechanisms are bronchoconstriction and, in certain instances, bronchorelaxation [15, 16], secretion from seromucous glands [17], and release of mediators (including prostaglandins and NO) from the airway epithelium. In the airways, the sensory neuropeptide CGRP is solely involved in the vasodilatation of bronchial arterioles in certain species [18]. However, CGRP has remarkable cardiac chronotropic and inotropic effects, and produces relevant coronary vasodilatation [19, 20]. Neurogenic inflammatory responses in the airways are substantially mediated by tachykinins and their receptors. In most species, apart for the rat, NK2 receptors mediate bronchoconstriction *in vitro* and *in vivo* [21, 22]. However, NK1 receptors appear to contribute to bronchoconstriction in the pig [23] guinea pig [24] and human small [25] and medium-size (Amadesi and Geppetti, unpublished observation) bronchi. NK1 receptors mediate the increase in airway blood flow, plasma extravasation and leukocyte adhesion in post-capillary venules [26–28], and secretion from seromucous glands in the ferret and man [17, 29]. NK1 receptor stimulation in the tracheal epithelium promotes the secretion of bronchorelaxant NO in the guinea pig [15] and prostaglandins in the rat and mouse [16]. NK3 receptors potentiate tachykinin-induced neurotransmission in postganglionic nerve terminals [22].

In the urinary tract, NK2 receptors mediate contraction of the smooth muscle of the urinary bladder, urethra and ureter. However, NK1 receptors also contribute to some of these responses [30]. Sensory CGRP plays a role in the motility of the urinary tract by producing relaxation in the ureter and in the bladder neck of the guinea pig [31]. Neurogenic inflammation heavily regulates the function of the iris sphincter muscle of the rabbit where sensory nerve stimulation and SP cause miosis. These same mechanisms seem to be present also in the pig [32], and there is some evidence of their presence in man [33]. At the gastrointestinal level, the source of SP/NKA and CGRP is dual. Together with extrinsic sensory nerve terminals, intrinsic neurons of the gut also express and release these neuropeptides. Thus, discrimination of the specific effects produced by neurogenic inflammation in this organ is more complex. Nevertheless, an important role is attributed to TRPV1-expressing neurons in the gastrointestinal tract in health and disease [34]. In cutaneous tissue, neurogenic inflammation includes mainly vascular effects. These are arterial vasodilatation (flare), plasma protein extravasation (wheal) and recruitment of inflammatory cells. Capsaicin-induced plasma extravasation has been clearly documented in the skin of rodents, but it is less evident in the human skin [35].

Modulation of neurogenic inflammation

Capsaicin and other pungent xenobiotics are unique activators of the TRPV1 channel [7], which is also stimulated by noxious heat (42°–53°C), low intracellular pH (pH 5–6) [36, 37] and certain lipid derivatives [38–40]. All these stimuli may thus promote neurogenic inflammatory responses. However, other agents able to 'sensitize' the TRPV1 may contribute to neurogenic inflammation. For instance, ethanol (and alcoholic beverages) reduces the threshold temperature for activation of TRPV1 by 8°C [41] and by this mechanism ethanol makes the TRPV1 excitable by the physiological temperature of 37°C. In this manner, ethanol may contribute to produce inflammatory responses in the airways [42] and in the dura mater (Trevisani and Geppetti, unpublished observations).

In addition to TRPV1, a plethora of channels and receptors are expressed in, and regulate, the activity of primary sensory neurons. Activation of inhibitory receptors on sensory nerves may limit neurogenic inflammatory responses [43, 44]. These receptors include neuropeptide Y, adenosine, serotonin 5-HT_{1D}, histamine H_3 and other receptors (see for review [44]). The list of agonist/receptor modulation is continuously increasing, as demonstrated by the recent evidence that also dopamine D_2 receptors [45] inhibit sensory neuron functioning. Agonists for these receptors may, thus, be considered as anti-inflammatory agents. Likewise, tachykinin receptor antagonists are regarded as potential anti-inflammatory drugs.

Stimulation of sensory nerves and the activation of nociceptive responses and neurogenic inflammation by a large series of agents has been described in a number of studies and review articles [10, 44, 46–48]. These stimuli include either autacoids like prostanoids, leukotrienes, histamine (H_1 receptors) and serotonin (5-HT_3 receptors) or changes in the milieu, like low extracellular pH, increased osmolarity and variations of the temperature, and inflammatory or tissue injury conditions like anaphylaxis [49].

Neurogenic arterial vasodilatation and plasma extravasation

Vasodilatation

In vivo studies directed at investigating neurogenic vasodilatation do not differ from those that investigate vasodilatation. The simple measurement of the reddened area of the flare in the human skin caused by stimulation of sensory nerve endings can be performed either directly with a ruler or by transferring the area of flare into an acetate sheet. The discovery that activation of sensory neurons results in cutaneous vasodilation, suggesting a neuronal 'efferent' function, was made more than 100 years ago using this relatively unsophisticated but still appropriate method [1, 50]. Technological improvement has offered novel methods for measuring arterial vasodilatation. One of the most successful has been the laser Doppler flowmetry, which uses the physical principles of laser technology and Doppler phenomenon to record variation in the fluid flow occurring in the underlying tissue. The major advantages of the method are that it is non-invasive and allows detection of the increase in blood flow with high sensitivity in well-defined and relatively small cutaneous areas. Laser Doppler flowmetry can also be applied to visceral organs, but its use for the study of neurogenic vasodilatation has been more oriented to the skin. For instance, the technique has allowed the definition of the contribution of complex mechanisms mediated by the sympathetic system to pain and neurogenic vasodilatation produced by capsaicin [51]. Finally, it is worth mentioning that laser Doppler flowmetry is particularly suitable for studying the effects and mechanisms of neurogenic vasodilatation in the human skin [35, 52, 53].

Another approach for the study of vascular neurogenic inflammation is intravital microscopy in the hamster cheek pouch. From the original description by Duling in 1973 [54], the method has received further modification [55]. Usually, the hamster is placed on a specially designed stage with a central depressed well, the right cheek is carefully everted and placed in the well and pinned on a silicon-rubber ring encircling the window. A single vascular layer is dissected out keeping an intact blood supply, and all connective tissue is removed. Drugs are administered though the cannulated jugular vein. The tissue is superfused with a warmed physiological solution. A selected area of microvasculature allows one (or more) arteriole and

Table 1 - Neuropeptides and receptors mediating major neurogenic inflammatory responses at the vascular and extravascular level

Function	Anatomical site	Mediator – receptor
Vascular neurogenic inflammation		
Arteriar vasodilation	Arterioles	CGRP - CRLR/RAMP
Plasma protein extravasation	Post-capillary venules	SP/NKA - NK1R
Leukocyte adhesion to the endothelium	Vanules	SP/NKA - NK1R
Extravascular neurogenic inflammation		
Smooth muscle contraction	Iris	SP/NKA - NK2/NK1-R
	Urinary bladder	SP/NKA - NK2/NK1-R
	Bronchus	SP/NKA - NK2/NK1-R
	Ureter/urethra	SP/NKA - NK2/NK1-R
Smooth muscle relaxation	Ureter	CGRP - CRLR/RAMP
	Bladder neck	CGRP - CRLR/RAMP
	Bronchus (NO/PGs epithelial release)	SP/NKA - NK1-R
Heart	Positive inotropic effect	CGRP - CRLR/RAMP
	Positive chronotropic effect	CGRP - CRLR/RAMP
Trachea/bronchus	Gland secretion	SP/NKA - NK1-R
	Increased mucosal permeability	SP/NKA - NK1-R

adjacent venule (each between 20 and 40 μm in diameter) to be viewed and monitored concomitantly in the same cheek pouch. The microvessels are viewed with a microscope and diameters of microvessels are usually quantified with an imaging system or more simply by a ruler.

Plasma protein extravasation

In physiological conditions, the microvascular endothelium forms a closed barrier between blood and tissues assuring exchange of oxygen, nutrients and hormones mainly involving channels and transporters proteins [56]. However, when a healthy post-capillary venule is exposed to a vasoactive agent, small pericellular gaps are transiently formed allowing the extravasation of macromolecules (for the majority plasma proteins) that remained imprisoned in the tissue [57–60]. SP and NKA released from unmyelinated sensory fibers around blood vessels, induce a rapid (<20

s) and transient (5–10 min) increase in permeability and the consequent plasma leakage (neurogenic mechanism). In contrast, non-neurogenic plasma extravasation can be induced by substances such as 5-HT, prostanoids, vascular endothelial growth factor (VEGF), histamine, antigen challenge, urotensin II and other stimuli [61–65], which directly alter blood vessel permeability.

Experimental quantitative analysis of the extravasated plasma proteins (PPE) can be made by diverse analytic methods. To date, the intravenous administration of the Evans blue dye (EB, 30–50 mg/kg) and the subsequent spectrophotometrical analysis of the dye is still the most common technique that allows the quantification of plasma extravasation [66]. Vasoactive stimuli are generally administered immediately after the EB injection. At 5–60 min (or in certain cases for more prolonged times) after dye administration, animals are transcardially perfused to eliminate all the circulating blood. Extravasated EB is extracted by incubation of the tissues in formamide and spectrophotometrically quantified at 620 nm. Evans blue content is calculated by linear regression analysis from external standards prepared in formamide. A relatively simple, reliable and sensitive method to measure PPE utilizes radioactive ^{125}I-labeled bovine serum albumin (^{125}I-BSA) administered intravenously [67]. In line with this model, vasoactive chemicals are administered after an intravenous injection of ^{125}I-BSA. After the transcardial perfusion, tissues are dissected, weighed, and counted for 20 min in a gamma counter.

An additional quantitative method for measuring the time course, the extent and location of PPE was recently established [68] in the rat skin. This model is based on video digital image processing. Briefly, after intravenous infusion of EB and the induction of PPE, the change in reflectance of the skin is recorded with a monochrome video camera. When EB is extravasated, the skin color changes to a dark blue, and the reflectance of the skin is, thus, greatly diminished. Images are digitally analyzed with software and the change in pixel intensity is determined in a precise and selected tissue region. The light intensity is than measured on a pixel by pixel basis in the digitized images.

Permeability of the tracheal mucosa

Exogenous tachykinins may also increase mucosal permeability in the guinea pig trachea [69, 70]. Exposure to ozone [69] and cigarette smoke [70] also increased tracheal permeability *via* a neurogenic inflammatory mechanism, as the effect produced by these two stimuli was abolished by NK2 receptor antagonism. Permeability of the tracheal mucosa has been measured by monitoring the appearance in the blood of horseradish peroxidase (HRP) previously instilled into the isolated tracheal segment [69, 70].

The original method [69, 70] consists of the exposure to the stimulus acting on sensory nerves (e.g., cigarette smoke) of anesthetized guinea pigs. A catheter is inserted into the femoral vein for drawing blood. A tracheal segment is, then, isolated *in vivo* between two polyethylene cannulae that are inserted by making tracheostomies. At 20 min after the exposure to the stimulus, HRP solution is slowly instilled to fill the lumen of the isolated tracheal segment. Blood samples are drawn before and following (at defined time intervals) the instillation of HRP *via* the catheter in the femoral vein. Plasma HRP levels are measured by an ELISA.

Inflammatory cell recruitment

Another key feature of neurogenic inflammation that invariably accompanies plasma protein extravasation is the adhesion of leukocytes (neutrophils and eosinophils) to the venule endothelium, the infiltration of the cells into the tissue and their return into the blood stream. An elegant way to monitor this important aspect of neurogenic inflammation utilizes, in the anesthetized animal, whole mount preparations of the tracheal wall where, after fixation of the tissue, inflammatory cells are identified and counted by conventional microscopy [26, 71]. The contribution of resident inflammatory cells, including mast cells, to neurogenic inflammation has also been proposed in the vessels of the dura mater [72]. Microscopical examination of the tissues coupled to pharmacological manipulation led to the conclusion that degranulated mast cells participate in building up the neurogenic inflammatory response.

Mucus secretion

Inhibition of airway neurogenic mucus secretion is of interest because neural mechanisms and mucus hypersecretion are implicated in the pathophysiology of certain severe respiratory conditions, such as asthma and chronic obstructive pulmonary disease (COPD). The chemical and physical properties of airway mucus are attributed to high molecular weight mucins. Mucins are secreted by specialized cells in the airways epithelium and submucosa (goblet cells, ciliated cells, mucous cells of the submucosal glands and Clara cells) [73]. Numerous detecting assays are available to study airway mucus secretion, using a variety of mucus markers [74–76].

In vitro systems using cultures of cells or explants are often used, but do not allow examination of the effect of nerve activities. *In vitro* nerve stimulation can be studied in isolated submucosal glands or in airway segments mounted in Ussing-type chambers. Radiolabeling of newly synthesized mucus is then studied through the use of radioactive tracers [77]. Nerve stimulation and the influence of reflexes can be studied in *in vivo* preparations in anesthetized animals. Tissue sections stained with Alcian blue and periodic acid-Schiff can be a useful methods to analyze the amount of intracellular mucin present. Alternatively, the mucins in collected secretions are assessed directly in binding assays using mucin antibodies or lectins (ELISA or ELLA) [78]. Antibodies against synthetic peptides derived from specific mucin genes have been prepared of different species [79, 80], and they will surely represent a clear improvement over existing assay systems.

There are many *in vivo* preparations for the study of mucus secretion. Preparations where mucus secretion is monitored over short periods (minutes to hours) are the predominant methods, and are carried out in anesthetized animals. In the hillocks technique [75], a segment of the upper trachea is incised ventrally in the mid line and transversely across both ends of the midline incision. Each midline cut edge is pulled back to expose the tracheal mucosal surface. Secretions are 'blotted' away with absorbent tissue and the exposed epithelial surface is sprayed with powdered tantalum, an inert metal, to give a uniform coating. The metal dust prevents the normal ciliary dispersion of secretions from the openings of the submucosal gland ducts which allows accumulated secretions to form elevations or 'hillocks' in the tantalum layer. Using video microscopy, the change in number of hillocks (usually an increase) in unit time (usually recorded over a number of seconds or minutes) before and after physiological intervention or drug administration is used as an index of a change in rate of secretion.

Micropipette sampling from individual submucosal glands [81, 82] is similar to the hillocks technique and involves exposure of a segment of upper tracheal mucosa. Superfused tracheal segments have been used extensively to investigate acute airway mucus secretion. Because of its size and amount of secretion produced, the cat has been the principal animal of choice [83], although the goose, ferret, rat and guinea pig have been also used successfully [83]. The tracheal segment is filled with physiological saline solution, which is collected at unit time intervals (minutes), after which the segment is refilled. A number of collections can be made over a period of a few hours. Mucus secretion into the collected liquid can be assessed in a variety of ways. Originally this was done by filling the trachea with oil and connecting one of the cannulae to a graduated tube. Passage of oil up the tube was used as an index of increased secretion. A radiolabel(s) is given into the tracheal segment at the beginning of the experiment and is incorporated over 2–3 h into newly synthesized mucins; the radiolabel is secreted into the tracheal segment. Dialysis of the collections removes unbound radiolabel and the amount of remaining mucin-bound label is quantified by liquid scintillation spectroscopy. Periodic acid-Schiff staining of the collected liquid followed by colorimetry is a useful adjunct to radiolabeling [84].

In contrast to the acute sampling methods, there are few techniques which allow long-term sampling of airway secretions [83]. These techniques involve either no anesthesia or anesthesia during establishment of the sampling system followed by recovery. For additional models for measuring the mucus secretion, please refer to the review [83].

Motor effects in the airways and urinary tract

The airways

As mentioned before, motor responses attributable to neurogenic inflammation are quite variable in the different tissues and according to the diverse species. Tachykinins, *via* NK2 and also, in minor part, NK1 receptors, cause bronchoconstriction in several mammals, with the exclusion of the rat and mouse [85]. For the study of bronchoconstriction *in vivo*, the most widely used model is the guinea pig where endogenously released tachykinins produced a remarkable narrowing of the tracheobronchial lumen. However, whereas in man NKA is one of the more potent spasmogenic agent *in vivo* and *in vitro* [86], there is no convincing evidence that capsaicin, which acts by releasing endogenous tachykinins, contracts the isolated human bronchus or causes bronchoconstriction *in vivo*. In general terms, there is little evidence in man of tachykinin-mediated neurogenic inflammation, whereas arterial dilatation caused by sensory CGRP seems to play a major role in human physiology and pathophysiology, as indicated by the recent observation that a CGRP antagonist was effective in the acute treatment of the migraine attack [87]. Bronchoconstriction, including bronchoconstriction induced by tachykinins released from sensory nerve endings, can be studied in rodents by the conventional methods adopted to investigate bronchoconstriction either indirectly by measuring the pulmonary insufflation pressure or other indices of bronchoconstriction [88], or more precisely by the measure of total pulmonary resistance [24]. Analysis of the bronchoalveolar lavage fluid based on the injection into the trachea, followed by recovery of an isosmotic amount of fluid allows the count of cells accumulated in the lung during pathophysiological events including neurogenic inflammation [89, 90].

The urinary tract

The role of neurogenic inflammation is of primary importance in the genitourinary tract [91]. Tachykinins released locally from sensory nerve endings in the urinary tract stimulate smooth muscle tone, contributing to both homeostatic and inflammatory motility of the ureter, bladder and urethra. They also trigger local and spinal reflexes aimed at regulating the functions of the organ in emergency conditions. In particular, tachykinin-mediated contractions contribute to bladder emptying and coordinated movements from the ureter to the urethra favoring the storage and expulsion of urines. In ability of CGRP to relax the ureter and the bladder also contributes to the physiological functioning of the urinary system [24]. Methods for the study of motility produced by neurogenic inflammatory mechanism of the urinary tract are performed predominantly *in vitro*. However, the study of cystometograms *in vivo* [30, 92] may offer peculiar information on the role of neurogenic inflammation. Measurement of blood flow and plasma protein extravasation have been extensively investigated in the urinary tract.

Acknowledgments

This review was supported in part by MIUR, Rome and by Fondazione Cassa di Risparmio di Firenze.

References

1   Lewis T (1927) *The Blood Vessels of the Human Skin and Their Responses.* Shaw and Sons, London

2   Jancso N, Jancso-Gabor A, Szolcsanyi J (1968) The role of sensory nerve endings in neurogenic inflammation induced in human skin and in the eye and paw of the rat. *Br J Pharmacol Chemother* 33: 32–41

3   Jancsó N, Jancsó-Gábor A, Szolcsányi J (1968) The role of sensory nerve endings in neurogenic inflammation induced in human skin and in the eye and paw of the rat. *Br J Pharmacol* 32: 32–41

4   Colpaert FC, Donnerer J, Lembeck F (1983) Effects of capsaicin on inflammation and on the substance P content of nervous tissues in rats with adjuvant arthritis. *Life Sci* 32: 1827–1834

5   Winter J, Forbes CA, Sternberg J, Lindsay RM (1988) Nerve growth factor (NGF) regulates adult rat cultured dorsal root ganglion neuron responses to the excitotoxin capsaicin. *Neuron* 1: 973–981

6   Clapham DE (1997) TRP channels as cellular sensors. *Nature* 426: 517–524

7   Caterina MJ, Schumacher MA, Tominaga M, Rosen TA, Levine JD, Julius D (1997) The capsaicin receptor: a heat-activated ion channel in the pain pathway. *Nature* 389: 816–824

8   Szolcsanyi J (1984) *Capsaicin-sensitive chemoceptive neural system with dual sensory-efferent function.* Akademiai Kiado, Budapest

9   Szallasi A, Blumberg PM (1999) Vanilloid (capsaicin) receptors and mechanisms. *Pharmacol Rev* 51: 159–212

10   Geppetti P, Holzer P (1996) *Neurogenic inflammation.* CRC Press, Boca Raton

11   Amara SG, Arriza JL, Leff SE, Swanson LW, Evans RM, Rosenfeld MG (1985) Expression in brain of a messenger RNA encoding a novel neuropeptide homologous to calcitonin gene-related peptide. *Science* 229: 1094–1097

12   Nakanishi S (1987) Substance P precursor and kininogen: their structures, gene organizations, and regulation. *Physiol Rev* 67: 1117–1142

13   Regoli D, Boudon A, Fauchere JL (1994) Receptors and antagonists for substance P and related peptides. *Pharmacol Rev* 46: 551–599

14   Foord SM, Wise A, Brown J, Main MJ, Fraser NJ (1999) The N-terminus of RAMPs is

a critical determinant of the glycosylation state and ligand binding of calcitonin receptor-like receptor. *Biochem Soc Trans* 27: 535–539

15 Figini M, Emanueli C, Bertrand C, Javdan P, Geppetti P (1996) Evidence that tachykinins relax the guinea-pig trachea *via* nitric oxide release and by stimulation of a septide-insensitive NK_1 receptor. *Br J Pharmacol* 115: 128–132

16 Frossard N, Rhoden KJ, Barnes PJ (1989) Influence of epithelium on guinea pig airway responses to tachykinins: role of endopeptidase and cyclooxygenase. *J Pharmacol Exp Ther* 248: 292–298

17 Geppetti P, Betrand C, Bacci E, Huber O, Nadel JA (1993) Characterization of tachykinin receptors in the ferret trachea by peptide agonists and non-peptide antagonists. *Am J Physiol* 265: L164–L169

18 Lundberg JM (1995) Tachykinins, sensory nerves, and asthma – an overview. *Can J Physiol Pharmacol* 73: 908–914

19 Franco-Cereceda A (1991) Calcitonin gene-related peptide and human epicardial coronary arteries: presence, release and vasodilator effects. *Br J Pharmacol* 102: 506–510

20 Franco-Cereceda A, Lundberg JM, Saria A, Schreibmayer W, Tritthart HA (1988) Calcitonin gene-related peptide: release by capsaicin and prolongation of the action potential in the guinea-pig heart. *Acta Physiol Scand* 132: 181–190

21 Advenier C, Naline E, Toty L, Bakdach H, Emonds-Alt X, Vilain P, Breliere JC, Le Fur G (1992) Effects on the isolated human bronchus of SR 48968, a potent and selective nonpeptide antagonist of the neurokinin A (NK2) receptors. *Am Rev Respir Dis* 146: 1177–1181

22 Advenier C, Lagente V, Boichot E (1997) The role of tachykinin receptor antagonists in the prevention of bronchial hyperresponsiveness, airway inflammation and cough. *Eur Respir J* 10: 1892–1906

23 Scheldrick RLG, Ball DI, Coleman RA (1990) Characterization of the neurokinin receptor mediating contraction of isolated tracheal preparations from a variety of species. *Agents Actions* (Suppl) 31: 205–210

24 Bertrand C, Nadel JA, Graf PD, Geppetti P (1993) Capsaicin increases airflow resistance in guinea pigs *in vivo* by activating both NK_2 and NK_1 tachykinin receptors. *Am Rev Respir Dis* 148: 909–914

25 Naline E, Molimard M, Regoli D, Emonds-Alt X, Bellamy JF, Advenier C (1996) Evidence for functional tachykinin NK1 receptors on human isolated small bronchi. *Am J Physiol* 271: L763–767

26 Baluk P, Bertrand C, Geppetti P, McDonald DM, Nadel JA (1995) NK1 receptors mediate leukocyte adhesion in neurogenic inflammation in the rat trachea. *Am J Physiol* 268: L263–L269

27 Corboz MR, Rivelli MA, Ramos SI, Rizzo CA, Hey JA (1998) Tachykinin NK1 receptor-mediated vasorelaxation in human pulmonary arteries. *Eur J Pharmacol* 350: R1–3

28 Piedimonte G, Hoffman JI, Husseini WK, Snider RM, Desai MC, Nadel JA (1993) NK1 receptors mediate neurogenic inflammatory increase in blood flow in rat airways. *J Appl Physiol* 74: 2462–2468

29   Rogers DF, Aursudkij B, Barnes PJ (1989) Effects of tachykinins on mucus secretion in human bronchi *in vitro. Eur J Pharmacol* 174: 283–286

30   Lecci A, Maggi CA (2001) Tachykinins as modulators of the micturition reflex in the central and peripheral nervous system. *Regul Pept* 101: 1–18

31   Maggi CA, Giuliani S, Del Bianco E, Geppetti P, Theodorsson E, Santicioli P (1992) Calcitonin gene-related peptide in the regulation of urinary tract motility. *Ann NY Acad Sci* 657: 328–343

32   Geppetti P, Patacchini R, Cecconi R, Tramontana M, Meini S, Romani A, Nardi M, Maggi CA (1990) Effects of capsaicin, tachykinins, calcitonin gene-related peptide and bradykinin in the pig iris sphincter muscle. *Naunyn Schmiedebergs Arch Pharmacol* 341: 301–307

33   Sicuteri F, Fanciullacci M, Nicolodi M, Geppetti P, Fusco BM, Marabini S, Alessandri M, Campagnolo V (1990) Substance P theory: a unique focus on the painful and painless phenomena of cluster headache. *Headache* 30: 69–79

34   Geppetti P, Trevisani M (2004) Activation and sensitisation of the vanilloid receptor: role in gastrointestinal inflammation and function. *Br J Pharmacol* 141: 1313–1320

35   Schmelz M, Petersen LJ (2001) Neurogenic inflammation in human and rodent skin. *News Physiol Sci* 16: 33–37

36   Tominaga M, Caterina MJ, Malmberg AB, Rosen TA, Gilbert H, Skinner K, Raumann BE, Basbaum AI, Julius D (1998) The cloned capsaicin receptor integrates multiple pain-producing stimuli. *Neuron* 21: 531–543

37   Bevan S, Geppetti P (1994) Protons, small stimulants of capsaicin-sensitive sensory nerves. *Trends Neurosci* 17: 509–512

38   Huang SM, Bisogno T, Trevisani M, Al-Hayani A, De Petrocellis L, Fezza F, Tognetto M, Petros TJ, Krey JF, Chu CJ et al (2002) An endogenous capsaicin-like substance with high potency at recombinant and native vanilloid VR1 receptors. *Proc Natl Acad Sci USA* 99: 8400–8405

39   Hwang SW, Cho H, Kwak J, Lee SY, Kang CJ, Jung J, Cho S, Min KH, Suh YG, Kim D et al (2000) Direct activation of capsaicin receptors by products of lipoxygenases: endogenous capsaicin-like substances. *Proc Natl Acad Sci USA* 97: 6155–6160

40   Zygmunt PM, Petersson J, Andersson DA, Chuang H, Sorgard M, Di Marzo V, Julius D, Hogestatt ED (1999) Vanilloid receptors on sensory nerves mediate the vasodilator action of anandamide. *Nature* 400: 452–457

41   Trevisani M, Smart D, Gunthorpe MJ, Tognetto M, Barbieri M, Campi B, Amadesi S, Gray J, Jerman JC, Brough SJ et al (2002) Ethanol elicits and potentiates nociceptor responses *via* the vanilloid receptor-1. *Nat Neurosci* 5: 546–551

42   Trevisani M, Gazzieri D, Benvenuti F, Campi B, Dinh QT, Groneberg DA, Rigoni M, Emonds-Alt X, Creminon C, Fischer A et al (2004) Ethanol causes inflammation in the airways by a neurogenic and TRPV1-dependent mechanism. *J Pharmacol Exp Ther* 309: 1167–1173

43   Maggi CA, Meli A (1988) The sensory-efferent function of capsaicin-sensitive sensory neurons. *Gen Pharmacol* 19: 1–43

44 Maggi CA (1991) The pharmacology of the efferent function of sensory nerves. *J Auton Pharmacol* 11: 173–208

45 Peiser C, Trevisani M, Groneberg DA, Dinh QT, Lencer D, Amadesi S, Maggiore B, Harrison S, Geppetti P, Fischer A (2005) Dopamine type 2 receptor expression and function in rodent sensory neurons projecting to the airways. *Am J Physiol Lung Cell Mol Physiol* 289: L153–158

46 Geppetti P (1993) Sensory neuropeptide release by bradykinin: mechanisms and pathophysiological implications. *Regul Pept* 47: 1–23

47 Holzer P (1988) Local effector functions of capsaicin-sensitive sensory nerves endings: involvement of tachykinins, calcitonin gene-related peptide and other neuropeptides. *Neuroscience* 24: 739–768

48 Holzer P (1991) Capsaicin: cellular targets, mechanisms of actions, and selectivity for thin sensory neurons. *Pharmacol Rev* 43: 143–201

49 Bertrand C, Geppetti P (1996) Tachykinin and kinin receptor antagonists: therapeutic perspectives in allergic diease. *Trends Pharmacol Sci* 17: 255–259

50 Bayliss W (1901) On the origin from the spinal cord of the vaso-dilator fibres of the hindlimb, and on the nature of these fibers. *J Physiol* 32: 1025–1043

51 Lin Q, Wu J, Willis WD (1999) Dorsal root reflexes and cutaneous neurogenic inflammation after intradermal injection of capsaicin in rats. *J Neurophysiol* 82: 2602–2611

52 Schulte H, Sollevi A, Segerdahl M (2004) The distribution of hyperaemia induced by skin burn injury is not correlated with the development of secondary punctate hyperalgesia. *J Pain* 5: 212–217

53 Steinhoff M, Stander S, Seeliger S, Ansel JC, Schmelz M, Luger T (2003) Modern aspects of cutaneous neurogenic inflammation. *Arch Dermatol* 139: 1479–1488

54 Duling BR (1973) The preparation and use of the hamster cheek pouch for studies of the microcirculation. *Microvasc Res* 5: 423–429

55 Hall JM, Brain SD (1994) Inhibition by SR 140333 of NK1 tachykinin receptor-evoked, nitric oxide-dependent vasodilatation in the hamster cheek pouch microvasculature *in vivo*. *Br J Pharmacol* 113: 522–526

56 Predescu D, Palade GE (1993) Plasmalemmal vesicles represent the large pore system of continuous microvascular endothelium. *Am J Physiol* 265: H725–733

57 Kenins P, Hurley JV, Bell C (1984) The role of substance P in the axon reflex in the rat. *Br J Dermatol* 111: 551–559

58 Majno G, Palade GE (1961) Studies on inflammation. 1. The effect of histamine and serotonin on vascular permeability: an electron microscopic study. *J Biophys Biochem Cytol* 11: 571–605

59 Majno G, Palade GE, Schoefl GI (1961) Studies on inflammation. II. The site of action of histamine and serotonin along the vascular tree: a topographic study. *J Biophys Biochem Cytol* 11: 607–626

60 Miller FN, Sims DE (1986) Contractile elements in the regulation of macromolecular permeability. *Fed Proc* 45: 84–88

61 Vergura R, Camarda V, Rizzi A, Spagnol M, Guerrini R, Calo G, Salvadori S, Regoli D

(2004) Urotensin II stimulates plasma extravasation in mice *via* UT receptor activation. *Naunyn Schmiedebergs Arch Pharmacol* 370: 347–352

62 Saria A, Lundberg JM (1983) Evans blue fluorescence: quantitative and morphological evaluation of vascular permeability in animal tissues. *J Neurosci Methods* 8: 41–49

63 Saria A, Lundberg JM, Skofitsch G, Lembeck F (1983) Vascular protein linkage in various tissue induced by substance P, capsaicin, bradykinin, serotonin, histamine and by antigen challenge. *Naunyn Schmiedebergs Arch Pharmacol* 324: 212–218

64 Gabbiani G, Badonnel MC, Majno G (1970) Intra-arterial injections of histamine, serotonin, or bradykinin: a topographic study of vascular leakage. *Proc Soc Exp Biol Med* 135: 447–452

65 Kondo S, Matsumoto T, Yokoyama Y, Ohmori I, Suzuki H (1995) The shortest isoform of human vascular endothelial growth factor/vascular permeability factor (VEGF/VPF121) produced by *Saccharomyces cerevisiae* promotes both angiogenesis and vascular permeability. *Biochim Biophys Acta* 1243: 195–202

66 Udaka K, Takeuchi Y, Movat HZ (1970) Simple method for quantitation of enhanced vascular permeability. *Proc Soc Exp Biol Med* 133: 1384–1387

67 Markowitz S, Saito K, Moskowitz MA (1987) Neurogenically mediated leakage of plasma protein occurs from blood vessels in dura mater but not brain. *J Neurosci* 7: 4129–4136

68 Gonzalez HL, Carmichael N, Dostrovsky JO, Charlton MP (2005) Evaluation of the time course of plasma extravasation in the skin by digital image analysis. *J Pain* 6: 681–688

69 Fu L, Kaneko T, Ikeda H, Nishiyama H, Suzuki S, Okubo T, Trevisani M, Geppetti P, Ishigatsubo Y (2002) Tachykinins *via* tachykinin NK(2) receptor activation mediate ozone-induced increase in the permeability of the tracheal mucosa in guinea-pigs. *Br J Pharmacol* 135: 1331–1335

70 Tagawa A, Kaneko T, Nishiyama H, Shinohara T, Sato T, Geppetti P, Ishigatsubo Y (2005) Cigarette smoke increases mucosal permeability in guinea pig trachea *via* tachykinin NK2 receptor activation. *Eur J Pharmacol* 507: 223–228

71 Baluk P, Bertrand C, Geppetti P, McDonald DM, Nadel JA (1996) NK1 receptor antagonist CP-99,994 inhibits cigarette smoke-induced neutrophil and eosinophil adhesion in rat tracheal venules. *Exp Lung Res* 22: 409–418

72 Markowitz S, Saito K, Buzzi MG, Moskowitz MA (1989) The development of neurogenic plasma extravasation in the rat dura mater does not depend upon the degranulation of mast cells. *Brain Res* 477: 157–165

73 Liu YC, Khawaja AM, Rogers DF (1998) Effects of the cysteinyl leukotriene receptor antagonists pranlukast and zafirlukast on tracheal mucus secretion in ovalbumin-sensitized guinea-pigs *in vitro*. *Br J Pharmacol* 124: 563–571

74 Davis CW, Abdullah LH (1997) *In vitro* models for airways mucin secretion. *Pulm Pharmacol Ther* 10: 145–155

75 Davis B, Roberts AM, Coleridge HM, Coleridge JC (1982) Reflex tracheal gland secretion evoked by stimulation of bronchial C-fibers in dogs. *J Appl Physiol* 53: 985–991

76   Davis B, Nadel JA (1980) New methods used to investigate the control of mucus secretion and ion transport in airways. *Environ Health Perspect* 35: 121–130

77   Gashi AA, Nadel JA, Basbaum CB (1987) Autoradiographic studies of the distribution of 35sulfate label in ferret trachea: effects of stimulation. *Exp Lung Res* 13: 83–96

78   Mazzuca M, Lhermitte M, Lafitte JJ, Roussel P (1982) Use of lectins for detection of glycoconjugates in the glandular cells of the human bronchial mucosa. *J Histochem Cytochem* 30: 956–966

79   Hovenberg HW, Davies JR, Carlstedt I (1996) Different mucins are produced by the surface epithelium and the submucosa in human trachea: identification of MUC5AC as a major mucin from the goblet cells. *Biochem J* 318: 319–324

80   Xing PX, Apostolopoulos V, Pietersz G, McKenzie IF (2001) Anti-mucin monoclonal antibodies. *Front Biosci* 1: D1284–1295

81   Ueki I, German VF, Nadel JA (1980) Micropipette measurement of airway submucosal gland secretion. Autonomic effects. *Am Rev Respir Dis* 121: 351–357

82   German VF, Corrales R, Ueki IF, Nadel JA (1982) Reflex stimulation of tracheal mucus gland secretion by gastric irritation in cats. *J Appl Physiol* 52: 1153–1155

83   Rogers DF (1997) *In vivo* preclinical test models for studying airway mucus secretion. *Pulm Pharmacol Ther* 10: 121–128

84   Somerville M, Richardson PS, Rutman A, Wilson R, Cole PJ (1991) Stimulation of secretion into human and feline airways by *Pseudomonas aeruginosa* proteases. *J Appl Physiol* 70: 2259–2267

85   Di Maria GU, Bellofiore S, Geppetti P (1998) Regulation of airway neurogenic inflammation by neutral endopeptidase. *Eur Respir J* 12: 1454–1462

86   Joos GF, Vincken W, Louis R, Schelfhout VJ, Wang JH, Shaw MJ, Cioppa GD, Pauwels RA (2004) Dual tachykinin NK1/NK2 antagonist DNK333 inhibits neurokinin A-induced bronchoconstriction in asthma patients. *Eur Respir J* 23: 76–81

87   Olesen J, Diener HC, Husstedt IW, Goadsby PJ, Hall D, Meier U, Pollentier S, Lesko LM (2004) Calcitonin gene-related peptide receptor antagonist BIBN 4096 BS for the acute treatment of migraine. *N Engl J Med* 350: 1104–1110

88   Ichinose M, Belvisi MG, Barnes PJ (1990) Bradykinin-induced bronchoconstriction in guinea pig *in vivo*: role of neural mechanisms. *J Pharmacol Exp Ther* 253: 594–599

89   Takebayashi T, Abraham J, Murthy GG, Lilly C, Rodger I, Shore SA (1998) Role of tachykinins in airway responses to ozone in rats. *J Appl Physiol* 85: 442–450

90   Su X, Camerer E, Hamilton JR, Coughlin SR, Matthay MA (2005) Protease-activated receptor-2 activation induces acute lung inflammation by neuropeptide-dependent mechanisms. *J Immunol* 175: 2598–2605

91   Candenas L, Lecci A, Pinto FM, Patak E, Maggi CA, Pennefather JN (2005) Tachykinins and tachykinin receptors: effects in the genitourinary tract. *Life Sci* 76: 835–862

92   Lecci A, Carini F, Tramontana M, Birder LA, de Groat WC, Santicioli P, Giuliani S, Maggi CA (2001) Urodynamic effects induced by intravesical capsaicin in rats and hamsters. *Auton Neurosci* 91: 37–46

Animal models of inflammatory bowel disease

Sreekant Murthy

Division of Gastroenterology and Hepatology and Office of Research, Drexel University College of Medicine, Philadelphia, USA

Introduction

Inflammatory bowel diseases (IBD) are genetically complex and multifactorial diseases of unknown etiology. They consist of two major illnesses: ulcerative colitis (UC) and Crohn's disease (CD). These are two of the most debilitating chronic diseases of the gastrointestinal tract, and lead to an unpredictable clinical course of multiple exacerbations and remissions of variable intensity. A wide range of symptoms may be present; however, clinically, the symptoms of UC, and particularly CD, may be overlapping (Tab. 1). A delay in diagnosis is common, as many of the symptoms do not suggest IBD initially because there are many other types of inflammatory bowel diseases that are not CD or UC. For instance, bacterial or parasitic infections of the bowel that cause inflammation are common at all ages; they are much more common than IBD and they are mostly curable. Even with specialized studies, it may be difficult to tell which type of IBD a person has. If this is the case, the diagnosis of "indeterminate colitis" is made with unclear diagnosis; therefore, it presents a challenge for the clinician to manage these patients [1]. The most serious complication associated with UC is the development of colorectal cancer. This requires frequent colonoscopy and testing, and long-term maintenance therapy with anti-inflammatory agents for preventing cancer.

In regards to onset, incidence and financial impact, about one third of all persons with IBD have the onset of their illness before adulthood. The peak age of onset is between 10–30 years and the disease persists for a large part of a person's life. Males and females are affected almost equally. There is a higher incidence in persons of Ashkenazic Jewish (Eastern European) ancestry. IBD tends to run in families. When one family member has IBD, there is a 15–30% chance that there is another affected family member. It is estimated that almost one million Americans are affected. The incidence of IBD varies from country to country; however, IBD has been increasing worldwide. As many as four million people worldwide suffer from a form of IBD. In the United States alone, IBD accounts for approximately 152 000 hospitalizations each year. The annual medical cost for the care of IBD patients in the Unit-

Table 1 - Clinical features of IBD

	Crohn's disease	**Ulcerative colitis**
Location	Entire GI tract	Colon
	Ileocolitis (40–55%)	Proctosigmoiditis (40–50%)
	Ileitis (30–40%)	Left-sided colitis (30–40%)
	Colitis (less common)	Pancolitis (20%)
		Backwash ileitis (<10%)
Clinical features	Abdominal pain	Abdominal pain
	Diarrhea	Bloody diarrhea
	Occult blood (less common)	Weight loss
	Low grade fever	Fever
Macroscopic features	Fibrotic and stenotic bowel	
	Strictures present	Strictures absent
	Deep, serpiginous and	Presence of punctate ulcers
	aphthous ulcers, Skip lesions,	
	"Cobble stone" appearance	
	Fissures and fistulas	Pseudopolyps
		Shortened colon
		Rectum always involved
	Anus is involved	Anus free of disease
Microscopic features	Transmural inflammation	Primarily mucosal
		Cryptitis, crypt abscess, dilated
		crypts, crypt branching and
		mucin depletion
	Focal cryptitis	
	Crypt abscess	
	Neuronal hyperplasia	
	Granulomas	Granulomas absent
Extraintestinal	Ankylosing spondylitis	Ankylosing spondylitis
manifestations	Psoriasis	Primary sclerosing cholangitis
	Iritis and uveitis	Pyoderma gangrenosum
Risk of cancer	Rare	High especially in pancolitis and
		long standing disease

ed States is considerable, estimated at over $2 billion. When adjusted for loss of productivity, the total economic cost is estimated to be nearly $2.6 billion [2].

IBD has been recognized for several decades, yet its etiology remains mysterious. Only in the last decade have substantial advances been made in defining the underlying mechanisms of chronic inflammation. IBD is now recognized as a multifactorial disease involving a number of overlapping genetic, environmental and immunological factors, working in concert in a dysregulated immune system in a genetically susceptible individual in the presence of altered enteric flora and a defective mucosal barrier, leading to chronic disease, as shown in Figure 1. In IBD, there are severe derangements in the structure and function of mucosal architecture, and increased presence of neutrophils and lymphocytes and other pro-inflammatory cells in the lamina propria. In addition, epithelial, endothelial, mesenchymal, and nerve cells display broad range of damage. It is now clear that hither to unsuspected effector and regulatory and immune-like functions abnormally interact with lymphoid cells to further aggravate the disease. Even acellular components such as extracellular matrix plays an important immunoregulatory activity under inflammatory conditions [3]. Recent studies have shown that at the cytokine level, CD versus UC is quite distinct in that CD is associated with overexpression of cytokines typical of the Th1 type, i.e., IL-12, TNF-α and IFN-γ, whereas UC is characterized by increased secretion of IL-5, probably no change in IL-4 and a lack of increase in IL-12 [4–6].

The conventional medical treatment of IBD consists of aminosalicylates, corticosteroids, immunosuppressive drugs (azathioprine, 6-MP, methotrexate, and cyclosporin), biological therapies, and antibiotics. The success with anti-TNF therapy, Infliximab, has propelled the development of several immunomodulating agents and anti-cytokine therapies in the hope that they may change the treatment strategy of IBD. These include, the humanized monoclonal antibody (mAb) CDP 571, the human soluble TNF p55 receptor antagonist (Onercept), the human mAb D2E7 (Adalimumab), the anti-human TNF antibody Fab' fragment-polyethylene glycol conjugate CDP 870, anti-CD3 antibody and the MAP-kinase inhibitor CNI-1493. In addition, agents that have been shown to suppress the adhesive interactions of inflammatory cells, such as a humanized mAb against α4 integrin (Natalizumab), have gone through trials for treating IBD [7]. However, until now, Infliximab is the only FDA-approved therapy for IBD, and it has shown great clinical remission of active CD. Despite those aggressive medical therapies, surgery is still required in the majority of patients.

Medical knowledge, treatment, and research that are involved in understanding the etiopathology of IBD have used laboratory animals for decades. The number of animal models used for preclinical studies and to understand the mechanisms of gastrointestinal inflammation are continually expanding. Currently, many animal models are being developed to focus on understanding the genetics and immunology of IBD. Many of these animal models are based on experimental needs; however, the positive impact of these animal models in advancing the medical knowledge of IBD cannot be underestimated. It is apparent that animal models do not entirely resemble human IBD, and also that one IBD patient does not entirely resemble another very well. Thus, the diversity of responses we observe in animal models is no dif-

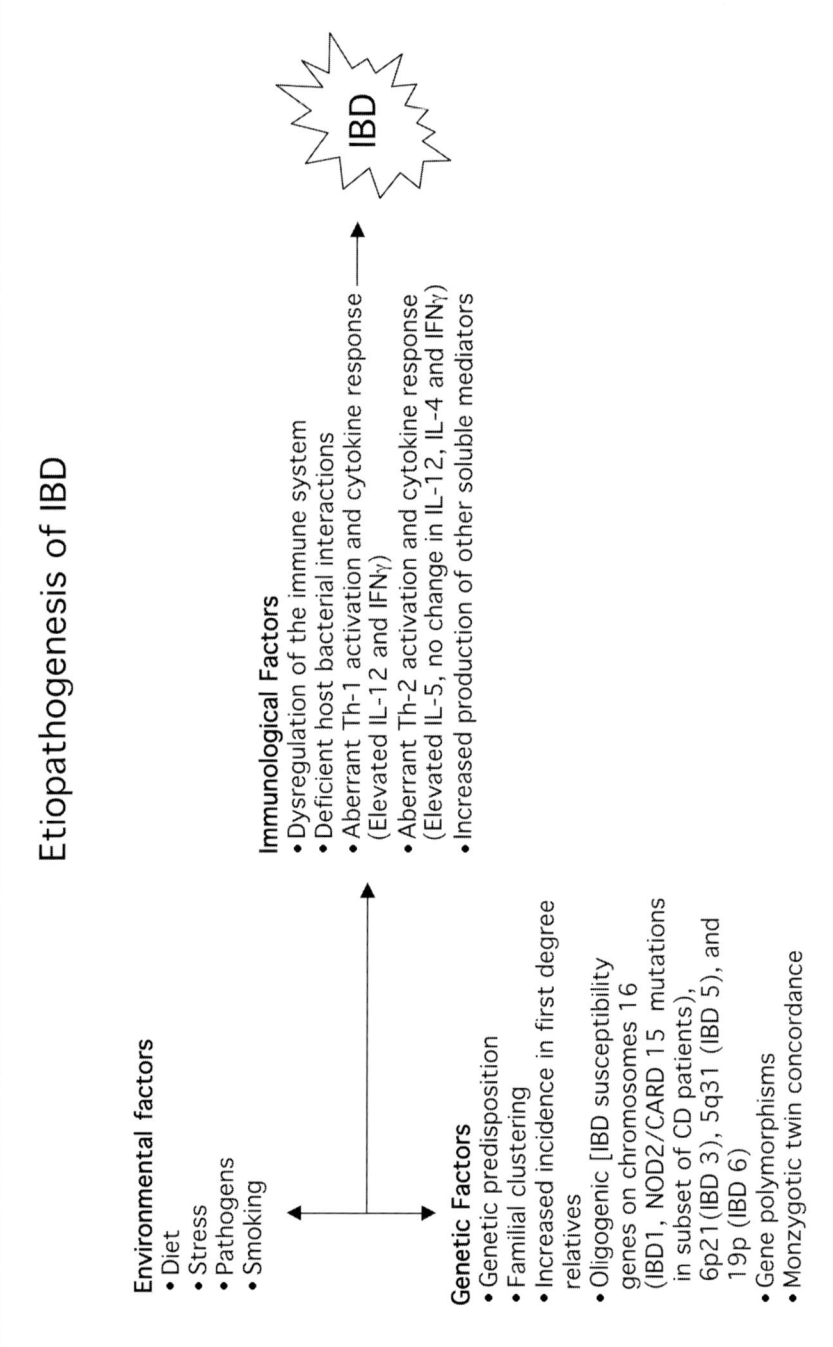

Figure 1
Current concept of the etiology of IBD.

ferent than what we observe in humans. Experience has been that even homogenously bred animals tend to respond differently to certain stimuli and treatments. The information provided in this chapter (Tab. 1) should facilitate selection of an appropriate model among the many species and strains, and the process for inducing intestinal inflammation, and gives the clinicopathological features and information on whether the model has been used for preclinical evaluation of drugs.

In general, these models can be divided into four major groups based upon how the disease is produced. These models include those that express intestinal inflammation spontaneously or naturally in their native or confined environment. The second category of the model includes those models in which intestinal inflammation can be induced by specific immunological or chemical agents that physically induce an immune response and mucosal damage in the local bacterial environment in relation to the damaging agent used. The third category includes those models that are genetically engineered by gene knockout, gene knockin or transgenic methods, so that the genetic factors related to the disease could be well defined and drugs that target a specific gene or cytokine can be tested. The last category includes "adoptive transfer models" in which experimental inflammation is induced by transferring certain T cell populations into an immune-compromised host lacking lymphoid tissue, such as SCID and Rag 2 knockout animals. In general, in most of these animal models, Th1 immune pathway is predominant which drives IL-12, TNF-α and IFN-γ production. Thus, there is an assortment of animal models of inflammation available for pharmacological testing of anti-inflammatory agents specific to IBD.

Spontaneous or natural models

As the etiology of IBD still remains unclear, natural models of intestinal inflammation continue to be of significant interest. These include animals that exhibit non-specific inflammation caused by infection, breeding or stress due to domestication, particularly in juvenile animals [8]. Since colitis observed in these animals is primarily due to bacterial infection, in some instances, the disease may be self limiting. The sporadic nature of the appearance of the disease, self-limiting inflammation, expense in procuring some of these animals and their husbandry contribute to their limited use in preclinical trials.

C3H/HeJBir mice

A substrain of C3H/HeJ mice has been reported that develops spontaneous cecitis and right-sided colitis in 60% of mice when they are 4–5 weeks old (Tab. 2). These mice are lipopolysaccharide (LPS) resistant due to a defect in the LPS locus on chromosome 4. A further substrain that develops spontaneous colitis with perianal

Table 2 - Animal models of inflammatory bowel disease

Model	Species	Method of induction	Time course	Disease location	Type of colitis	Successful therapies
I. Spontaneous models						
1. C3H/HeJBir	Mice	Natural	4–6 weeks	Cecum and colon	Acute	CRX-526 (TLR-4 antagonist) [10]
2. Cotton top tamarins	Monkeys	Natural	<1 year	Colon	Acute and chronic	Anti-α4β7, anti-TNF antibody, 5-ASA, 5-LO [9, 111–113]
3. SAMP1/YIT	Mice	Natural	30 weeks	Ileum	Chronic	TNF antibody [18], antibodies against anti-adhesion molecules [18, 114–117]
II. Immunological models						
1. Immune complex-mediated	Rabbits and rats	Injection of albumin anti-albumin complex + a formalin enema	3 h–8 weeks	Colon	Acute	5-ASA, IL-1-ra, anti-IL-8, beta integrin 2 antagonist, recombinant CSF, anti-a4 [22, 24, 118–122]
III. Bacteria and bacterial products models						
1. Helicobacter hepaticus	SCID mice	Infection	2–90 days	Colon	Chronic	–
2. Peptidoglycan poly-saccharide (PGPS)	Rats	Transmural injection of PGPS		Ileum	Chronic, relapsing	IL-10, IL-1ra [29, 30]
IV. Chemically-induced models						
1. Trinitrobenzene sulfonic acid (TNBS)	Rats, rabbits and mice	TNBS enema (20–30 mg in 30–50% EtOH)	3 days–8 weeks	Small intestine or colon	Acute and chronic	Steroids, 5-ASA, TNF-α, fish oil, Zileuton, MK 886, EGF, KGF, L-NAME, IL-10, IL-11, anti-IL-12 antibody, IL-1 receptor antibody, anti-IFN-γ, anti-IL-6 receptor antibody, Fib antisense [35, 123–136]

Model	Animal	Method/dose	Duration	Location	Acute/chronic	Therapeutics [references]
2. Dextran sulfate (DSS)	Hamsters, mice and rats	2–10% DSS feeding	5 days–15 weeks/longer	Colon	Acute and chronic	IL-1ra, TGF-β2, TNF-α, cyclosporin, anti-oxidants, FLAP inhibitor, IL-10, anti-ICAM, NF-κB antisense, ICAM 1 antisense, TNF-antisense, PPR gamma ligand and RDP58 (TNF inhibitor) [137–149]
3. Acetic acid enema	Rats	1–10% acetic acid enema (buffered or unbuffered)	1 day–3 weeks	Colon	Acute	5-ASA, L-NAME, IL-1ra, SOD inhibitors, PGE [150–153]
4. Carrageenan	Rats, guinea pigs and rabbits	Variable oral dosing	1–4 weeks	Cecum and colon	Acute and chronic	Metronidazole [154]
5. Indomethacin	Rats	Oral or s.c once or twice	<1–8 days	Small intestine	Acute	5-ASA, 5-LO and TXB-2 inhibitors [155–157]
6. Oxazalone	Mice and rats	Intracolonic	Rapid	colon	Acute	Cyclosporin, budesonide and anti-IL-4 antibody [42, 43]
V. Miscellaneous chemical models						
1. Peroxynitrite	Rats	Enema	< 1 day–3 weeks	Colon	Acute	L-NAME [158]
2. Mitomycin C [72]	Fischer rats	i.p. 2.75 mg/kg		Colon	Acute	–
3. Iodoacetamide [159]	Rats	3% enema		Colon	Acute and chronic	–
4. Phorbal-12-myristate-13-acetate (PMA) [160]	Rabbits	Enema	4 days	Colon	Acute	–
VI. Genetically engineered models						
1. HLA-B27/β2 microglobulin	Rats	Transgene	>10 weeks	Intestines	Acute and chronic	IL-11 [161]
2. STAT 4 transgenic mice		Systemic administration of DNP-KH limpet hemocyanin			Chronic	

143

Table 2 (continued)

Model	Species	Method of induction	Time course	Disease location	Type of colitis	Successful therapies
3. N-cadherin		Transgene		Colon	Chronic	
4. IL-2 knockout	Mice	Gene deletion	>5 weeks	Colon	Acute and chronic	Anti IL-12 antibody [162]
5. IL-10 knockout	Mice	Gene deletion	>6 weeks	Colon and small intestine	Acute and chronic	Anti-IL-12 antibody, anti-a4 [163, 164]
6. TCR knockout	Mice	Gene deletion	>30 weeks	Colon	Acute and chronic	Anti-IL-12 antibody (neu 71), anti-IL-6 receptor antibody (52 Neu), NF-κB antisense (neu 112) [123, 136, 163]
7. Gαi2 knockouts	Mice	Gene deletion	>15 weeks	Colon	Acute and chronic	Anti-IL-4 antibody (anti-IL-4 antibody, IL-1 receptor antibody) [42, 162, 165]
8. Mdr1a (multiple drug resistance gene)	Mice	Gene deletion	8–36 weeks	Colon	Acute and chronic	Antibiotics [91]
9. TNF-α 3'(69 bp) deletion	Mice	Deletion 3'-AU-rich region 589	3–8 weeks	Ileum	Acute and chronic	–
10. Keratin 8	Mice	Gene deletion		Colon	Acute and chronic	–
11. TGF-β1 knockouts	Mice	Gene deletion		Colon	Multiorgan, early death	–
12. IL-7 transgenic	Mice	Transgene	>8 weeks	Colon	Acute and chronic	–
13. A20	Mice	Gene deletion	<1 week	Intestines	Acute, early death	–
14. STAT 3	Mice	Deletion of STAT 3 gene in bone marrow cells	<2 months	Small intestine and colon	Acute and chronic	–

15. CD40	Mice	Transgene	8–15 weeks	Colon	Acute and chronic	–
VII. Adoptive transfer models						
1. CD45RB^hi	Rag2^-/- and SCID mice	Adaptive transfer of CD45RB^hi CD4+ T cells		Colon	Chronic	IL-10, anti-IFN-γ, anti-TNF-α anti-IL-6 receptor antibody [99, 136]
2. 2. Tgε 26	Mice	Transfer of syngeneic bone marrow		Colon	Chronic	Anti-TNF-α and IFN-γ [110]

– = Data not available

PG/PS, Peptidoglycan polysaccharide; TNBS, trinitrobenzene sulfonic acid; 5-ASA, 5-aminosalicylic acid; TNF, tumor necrosis factor; AA, arachidonic acid; EGF, epidermal growth factor; KGF, keratinocyte growth factor; ROM, reactive oxygen metabolites; TGF, transforming growth factor; IGF, insulin-like growth factor; IFN, interferon; IL, interleukin; PGE, prostaglandin E; SOD, superoxide dismutase; TXB, thromboxane B; NO, nitric oxide; L-NAME, L-arginine methyl ester; mAb, monoclonal antibody; 5-LO, 5-lipoxygenase; FLAP, 5-lipoxygenase activating protein; NF-κB, nuclear factor kappaB; PPR, peroxisome proliferator-activated receptor; ICAM, intercellular adhesion molecule; CSF, colony-stimulating factor; TLR, Toll-like Receptor. For brevity, only a few representative references are provided.

ulcers and occult blood loss in 98% of animals has also been reared and named C3H/HeJBir mice [9]. These mice show remarkable reactivity to enteropathogenic bacterial antigens and dextran sulfate, resulting in fulminant colitis. The model has been used in genetic mapping studies to identify genes that are susceptible or resistant to colitis. A lipid-mimetic synthetic CRX-526, which has antagonistic activity for TLR4 [10], has been to shown to inhibit inflammation in this model.

Cotton top tamarins

Sanguinas oedipus or cotton top tamarins (CTT) belonging to marmoset family are new world monkeys native to north central Columbia, South America. When these animals are in their natural environment, they show little or no disease. In a confined environment, they develop severe colitis resembling human UC, including heightened occurrence of colon cancer [11, 12]. Colitis in CTT is associated with anorexia, watery diarrhea or bulky soft stools, dehydration and weight loss. Histologically, acute colitis shows mucin depletion, an influx of polymorphonuclear leukocytes and crypt abscesses. Lymphocytes, plasma cells and eosinophils are seen, but gross ulcers are less common. In the chronic state, neutrophils and granulomas are not seen, but the lamina propria contains monocytes and plasma cells. Dysplasias are commonly seen in areas of chronic inflammation, suggesting that severe inflammation may predispose these animals to neoplastic changes. These animals also develop autoantibodies against epithelial autoantigens [13]. It appears that CTTs do not acclimate well and they are metabolically stressed at temperatures less than 32°C, which may initiate and perpetuate chronic colitis [14]. CTTs exhibit low levels of MHC class I polymorphisms or allelic diversity and, therefore, are highly vulnerable for viral attack and possible susceptibility to colitis [15]. Most importantly, CTTs respond very well to 5-aminosalicylic acid (5-ASA) and 5-lipoxygenase inhibitor therapies [11, 16]. The success of 5-ASA therapy has contributed to long-term remission and survival of CTTs in captivity. It also boosts confidence in the model's ability to impending success of novel therapies in human trials. The drawbacks for using this model are pancolitis at the onset of disease, limited availability of animals since there are only a few primate centers in the world, housing, expense and the endangered species status of the animals.

SAMP1/YIT mice

Kosiewicz et al. [17], in collaboration with Japanese investigators, described a mouse model of spontaneous colitis that has remarkable similarity to human CD. The model was derived by brother-sister mating of AKR strains that develop spontaneous ileitis. Inflammation is transmural, localized to the terminal ileum with heavy infiltration of activated CD4[+] and CD8α[+]TCRαβ[+] T cells into the lamina propria, and decrease in intraepithelial lymphocyte CD8α[+]TCRγδ[+]/CD8α[+]TCRαβ[+] ratio. Adaptive transfer of these cells to normal mice results in inflammation, which can be prevented by pretreatment of normal mice with neutralizing anti-TNF antibody [18]. The similarity with CD and response to preventive TNF therapy presents the opportunity to test compounds directed for treating CD.

Immunological, bacterial and bacterial byproduct model

Animal models of experimental inflammation induced by immunological and chemical agents comprise of a whole host of murine and other animals. Development and use of these models are one of the fundamental reasons for advancing our understanding of immunopathogenicity and drug discovery in IBD today. Many of these models whether induced by immunological agents or induced by physical damage, elicit a Th1-like T cell cytokine response, Th2 responses are also seen. In some cases [19], there may be Th1 and/or Th2 responses and there may even be a switch of Th1 to Th2 response. It is unclear why there is dominant Th1 response; however, the experimental evidence points to damaging agents aggravating the inflammatory response through antigens in the normal intestinal flora, causing transmural infiltration of inflammatory cells. The general assumption is that whenever a Th1 response is elicited, there is heavy involvement of transmural inflammatory cells despite the presence of mucosal damage presenting CD-like pathological features, and, in Th2 response, there is increased mucosal damage and increased influx of polymorphonuclear cells presenting UC-like features. Researchers interested in understanding the complex interplay of the immune system in these animal models are referred to an article published by Strober et al. [20].

Models produced by immunological agents

One of the earliest immunologically derived model is immune complex colitis produced in rabbits by a combination of the intravenous injection of human albumin and anti-albumin complex, followed by rectal administration of dilute formalin (Tab. 2) [21]. The purpose of using formalin is to damage the colon and render permeability of systemic albumin and anti-albumin complex into the damaged region to elaborate a local immune reaction. The soluble complex (0.5–0.75 ml) is given intravenously through the marginal ear vein (rabbits) or tail vein (rats) 2–3 h after a 0.4–4% formalin enema. There are three distinct methods described for producing colitis in rats and rabbits. The first method involves the injection of immune complex in non-sensitized animals and the second approach involves first sensitizing the animals with enterobacterial antigen of Kunin [22]. In the third, rats or rabbits are preimmunized with 100 mg *Escherichia coli* 014:K7:H-antigen emulsified in complete Freund's adjuvant before a second challenge with albumin-anti-albumin complex and formalin enema [23]. The inflammation resolves after 2 weeks, without clear evidence of chronic colitis except in animals that are presensitized, in which inflammation can last for several months. Histologically, chronic colitis exhibits many features of acute colitis with an increased influx of neutrophils within 24 h, followed by mixed inflammation, crypt abscesses, crypt distortion, ulceration and necrosis. The mechanism of injury, the mediators involved and results of drug efficacy trials are shown in Table 1. Colitis is associated with increased synthesis of arachidonic acid metabolites, which occurs a few hours after the activation of IL-1 gene expression [24]. Pretreatment and treatment with IL-1 receptor antagonist (IL-1ra) within the first 33 h after the induction of colitis diminishes inflammation scores and necrosis. Also, IL-10 has been shown to ameliorate acute colitis. Sulfasalazine treatment is effective [23, 25], but steroids do not work.

Bacteria and bacterial by product models

The postulation that an infectious agent is a causative factor of IBD has persisted for many decades. In IBD patients, the presence of many different bacterial agents has been documented; however, none of these pathogens have ever been directly implicated as primary etiopathic agents. Animal models have provided the most credible evidence for our understanding today that IBD may occur due to a loss of tolerance to resident bacteria. Animal model studies have unquestionably shown that a large pool of antigens is present in the lamina propria, but only some of these antigens are responsible for activating lymphocytes. Manipulation of animal models with combinations of broad spectrum antibiotics as opposed to a single antibiotic yields beneficial effects, suggesting that a host of bacteria are involved in the pathogenesis. The observation that some bacteria like *Lactobacillus* and *Bifidobacterium* species, when ingested into the host animal, show some beneficial effects has opened up possibilities for investigating the mechanisms of prebiotic and probiotic treatments.

Animal models also have shown that certain pathogenic bacteria when naturally present in immune incompetent mice also cause inflammation. Ward et al. [26] described that 5% of immunodeficient athymic Ncr-nu/nu, BALB/c AnNCr-nu/nu, C57BL/rNCr-nu/nu and C.B17/Icr-scid/ncr mice naturally infected with *Helicobacter hepaticus* develop typhlitis, colitis, proctitis and rectal prolapse (Tab. 2). *Chlamydia trachomatis* when rectally inoculated produce proctitis in Cynomolgus monkeys, *Macaca fasicularis* [27]. Proctitis is seen after 1–2 weeks and resolves rapidly within 2 weeks. Interestingly, *C. trichomatis* is not involved in human IBD.

Similar to certain pathogenic bacteria, bacterial cell wall products and endogenous antigens derived from ingested food material have been shown to be highly antigenic. The bacterial secretory product formyl-methyl-leucyl-phenylalanine (FMLP) and cell products, such as lipopolysaccharides and peptidoglycan polysaccharides (PG/PS), are highly antigenic, and they activate many possible cascades of inflammation and cause damage to the colonic mucosa and vascular tissue architecture. Sartor et al. [28] showed that a single subserosal injection of group A PG/PS results in ileal, cecal and distal colonic, chronic, spontaneously relapsing, transmural, granulomatous enterocolitis, arthritis, anemia and hepatic granulomas in genetically susceptible Lewis rats (Tab. 2). The acute phase is rapid and resolves after 2 weeks. Chronic granulomatous colitis emerges after 2 weeks; it reaches a peak at 3 weeks after injection. Buffalo rats and Fischer F344 rats, which are MHC-matched with Lewis rats, develop self-limiting inflammation without any signs of extraintestinal or systemic inflammatory reactions. The mechanism of injury seems to be related to increased mucosal permeability and myeloperoxidase activity, and nitrous oxide (NO) and collagen synthesis. Treatment of PG/PS-induced colitis with IL-1ra and IL-10 ameliorates inflammation in rats [29, 30]. The advantage of using this model is that it resembles CD by showing chronic, relapsing, transmural and granulomatous inflammation including extraintestinal manifestation and arthritis. The disadvantages are that it is species/strain specific, requires laparotomy and direct serosal injection of PG/PS and produces enterocolitis and inconsistent mucosal ulcerations. PG/PS is expensive and only available from a few commercial sources.

A close analogue of the PG/PS model of enterocolitis is the incomplete Freund's adjuvant (IFA)-induced colitis in rats and guinea pigs [31]. The method of producing injury is the same as in the PG/PS model where IFA mixed with heat-killed *Mycobacterium tuberculosis* prepared in mineral oil is injected intramurally into the subserosal regions to produce chronic colitis lasting for 4 weeks. The induction of disease requires surgical intervention. Preclinical trials have shown that 5-ASA and an inhibitor of inducible NO synthesis provide significant protection in this model.

Animal models of physical injury

Acetic acid model

Acetic acid-induced colonic injury, first introduced by MacPherson and Pfeiffer [32], requires abdominal incision and introduction of 5% acetic acid into the colonic lumen. The method has now been changed to introduce various concentrations of acid as an enema into rats, guinea pigs, rabbits and mice, with or without prepping or buffering the colon. The most commonly used concentration is 3–5% acetic acid given as an enema in volumes less than 1 ml (Tab. 1). At this concentration, animals develop physical injury at the site of delivery. Colitis is easy to produce and highly reproducible. Injury occurs rather rapidly with the initiation of injury to the mucosal surface altering the mucosal barrier. Peak damage, with severe acute inflammation, ischemia and erosions occurs within 24 h and resolves within 2 weeks with little or no chronic inflammation. The disease is characterized by the loss of surface epithelium, ulcerations, the loss of goblet cells, crypt abscesses, edema and an influx of neutrophils in the lamina propria, occurring within 2–3 days. The mechanism of injury is unclear, yet it is known that it involves a protonated form of this acid at pH 2–3 since buffered acid and HCl at the same pH do not cause epithelial injury [33]. Oxygen-derived free radicals are also suggested to be involved in producing injury. Colonic injury is associated with increased production of arachidonic acid metabolites produced by infiltrating neutrophils. Many therapies used for the treatment of human IBD, including some unconventional therapy with hitamine-1 antagonist and Mesoprostol, have been shown to inhibit inflammation in this model. However, the popularity of this model has diminished since the model shows absence of significant chronic inflammation and initial epithelial cell injury, producing a later immune activation that is opposite to that of human IBD.

Trinitrobenzene sulfonic acid models

Since it was first described in 1989 by Morris [34], trinitrobenzene sulfonic acid model has been very popular. Multitude of articles have been published to describe the use of this model for both experimental and therapeutic purposes. Its popularity is well founded because a single application of TNBS in rats, mice, guinea pigs, dogs and rabbits produce rapid, reliable and reproducible disease (Tab. 2) [34–38]. Since the model can be produced in multiple species, it renders the possibility of second or third species verification to rule out false negatives due to species-specific results. The induction of colitis by TNBS is simple, requiring an introduction of 20–80 mg TNBS dissolved in 30–50% ethanol as an enema in the rat colon, but the degree of disease and time required to produce the injury may vary between laboratories. TNBS induces peak acute inflammation within a week, which gradually progresses into chronic inflammation lasting for about 8 weeks. Certain mice are resistant to TNBS [30]. In TNBS-susceptible mice, acute injury can be produced by low concentrations of TNBS and ethanol and the injury can be visualized within 2–3 days. The disease may resolve rapidly compared to rats. Repeat enemas of TNBS may accentuate injury, but oral feeding of TNBS results in significant oral tolerance [39].

In both rats and mice, the disease shows some resemblance to CD with "skip lesions", "cobble stoning", linear ulcers and transmural inflammation. Rats show the presence of giant cell and granulomas. In SJL/J mice, TNBS-ethanol produces transmural inflammation with granulomas with severe weight loss and bowel wall thickening [35]. The mechanism of injury involves a trinitrophenyl group, which acts as a hapten by covalently bonding with cell surface proteins and presenting the MHC class II peptide aggregates to CD4+ T cells through antigen-presenting cells to induce a CD4[+] T cell immune response and cytokine production [35, 40]. Inflamed tissues express increased levels of IL-12 and IFN-γ. The elevated level of IFN-γ has also been suggested to activate macrophages to produce chemotactic factors and proinflammatory cytokines [35]. IL-12 and IFN-γ antibody-treated animals not only prevented the induction of TNBS-induced colitis [35], but IL-12 antibody was therapeutic in established disease [35]. Also, TNF-α mAb and rIL-4 down-regulate colitis. These data from studies done SJL/J mice suggest that TNBS colitis is mediated by a Th1 response, and that agents that modify this Th1 response help down-regulate colitis. Thus, many histopathological and cytokine profile observed in this model are reminiscent of the pattern of injury observed in human IBD.

In TNBS-induced rat colitis, the injury is associated with a substantial increase in myeloperoxidase levels, suggesting that neutrophils play a significant role in the production of acute injury. Over the last decade, this is by far one of the most popular models used for the preclinical evaluation of drugs that have been published. The success of these therapies provides valuable comparisons for future drugs to be tested (Tab. 1). The advantages of this model is in the simplicity and relative low cost of producing colitis. The disadvantages are that the reproducibility of the model is dependent upon the dose of TNBS, the lot of TNBS, the concentration of ethanol, and, most importantly, the genetic background and microflora in the animal facilities, as there are differences in the expression of disease in various species and the strain of animals used in various facilities.

Oxazalone model

Colitis induced in rats and mice by intrarectal administration of a hapten oxazalone (4-ethoxymethylene-2-phenyl-2-oxazolin-5-one), dissolved in carmellose sodium in peanut oil or in ethanol [41, 42], has been shown to have UC-like features with mucosal epithelial damage and ulceration. In the rat, 5-ASA, cyclosporin A and budesonide have shown to have therapeutic benefit. Interestingly, in SJL/J mice, the disease is Th2 related, characteristically producing greater mucosal damage, and overexpression of IL-4 and IL-5. Administration of IL-4 antibodies have a therapeutic benefit in this model. This model is easy to produce, and animals rapidly develop distal colitis. Although the mortality rate is high in this model, the surviving animals show several UC-like features making this model suitable unique for testing agents to treat UC.

Dextran sulfate model of colitis

In 1985, Ohkusa [43], and in 1990, Okayasu et al. [44], reported two new animal models of colitis in Syrian hamsters and BALB/c mice, respectively, by feeding them

3–10% dextran sulfate sodium solution (DSS) dissolved in drinking water (Tab. 1). In our laboratory, we further characterized the model by producing colitis in outbred female Swiss Webster mice [45, 46]. The disease was induced by feeding 5% DSS (MW 30–40 kDa, ICN Biochemicals) for 5 days to produce "acute" colitis and 7 days of DSS feeding, followed by 21 days of plain water to produce "chronic" colitis. In addition, long-term or cyclic administration of DSS for four cycles produces severe colitis, which after 3, 6 and 9 months results in the development of dysplasia or adenocarcinoma in mice, rats, hamsters. Pretreatment of DSS-fed mice with azoxymethane results in the development of colon cancer in the inflamed regions [47]. Published studies show that guinea pigs are more susceptible to DSS injury than any other species studied so far [48]. In our studies [45], we have shown that outbred mice develop lesions in the mid-colon first, followed by the distal colon. Fisher 344 rats [49], BALB/c [34] and CBA/j [44] mice develop left-sided colitis. Conversely, hamsters [50], guinea pigs [48] and Wistar rats [51] develop right-sided colitis. These studies suggest species, strain and site specific susceptibility of the colonic mucosa, which probably uncover certain genetic differences in susceptibility to DSS between and within species.

Evaluation of colitis in this model has utilized both functional scores and histology. Functional score utilizes a method of determining a disease activity index based on daily measurement of body weight, stool consistency, and testing for occult blood in the stool or bleeding per rectum [45, 46]. The disease in outbred mice begins to develop 3 days after feeding DSS, reaching peak disease on day 7. The most severe disease seen on day 7 is not ideal for preclinical testing since some of the animals become wasted with severe disease. The disease activity index shows an excellent correlation with crypt architectural changes [45]. The disease spares the small intestine. Histopathological changes, which predate clinical symptoms, include a systematic progressive non-inflammatory crypt dropout, hyalination in the lamina propria and in the pericrypt regions. The earliest changes are focal, involving less than 10% of the mucosal surface and involving two to three crypts. After 4 days of DSS, histological changes become more confluent and diffuse, with evidence of significant loss of crypts involving 15–75% of the surface area. There are minimal, mixed inflammatory infiltrates of which granulocytes are more abundant. Monocytes and macrophages are rarely seen at this time. After 5 days of DSS, pathological changes become more confluent with the loss of surface epithelium, the formation of erosions, and the emergence of an early hyperplastic epithelium. Inflammation is florid, with granulocytes, lymphocytes and plasma cells, and a minimal presence of monocytes and macrophages. Some cryptitis is seen, but crypt abscess are not present. Once DSS feeding is stopped, the animals progress to developing severe chronic inflammation, and subsequent dysplasia is seen in some animals. Chronic inflammatory infiltrates seen at this point consist of only a few granulocytes, an abundance of lymphocytes, plasma cells, monocytes and macrophages. Many of the histological changes seen at this stage are reminiscent of human chronic UC. Even 60 days after discontinuing DSS, chronic inflammation and erosions are still present, providing a large enough window for therapeutic intervention without the inherent risk of self healing. Continuous, cyclic or even one-time administration of DSS results in the development of dysplasia after 3 months. The incidence of dysplasia may vary by the way DSS is administered, but is usually around 20%. DSS-fed Min mice show an early development of colon cancer [52].

The mechanism by which DSS induces colonic damage is beginning to emerge. The degree of sulfation does not seem to be a contributing factor for disease induction. However, since colonic bacteria can dissociate sulfate from DSS, the free sulfate present in the intestine could act as a substrate to produce H2S, which could significantly interfere with cellular metabolism to induce a toxic effect on the epithelium [53]. Other possible mechanisms include alterations in luminal bacterial ecology and the activation of monocytes, macrophages and mast cells. Germ-free animals [54] produce colitis similar to that observed in non-germ free animals. Thus, bacteria are not primarily involved in producing disease. However, broad spectrum antibiotics have provided prophylactic or therapeutic benefit [50, 55–58] making the involvement of bacteria controversial. Mice lacking TLR-2 and TLR-4 receptors were less susceptible to DSS compared MyD88 (myeloid differentiation factor 88) knockout animals, suggesting that this pathway is important for protecting animals from infection and DSS-induced tissue damage [59]. DSS induces inflammation in athymic nude mice, severe combined immune deficient mice (SCID) [60] and antibody-dependent helper T cell-, B cell- and NK cell-depleted mice [61], indicating that lymphocytes are not involved in the disease induction process. The findings that DSS induces colitis in Rag 2-deficient mice [61] is indicative of the activation of macrophages [62, 63]. In the DSS model, the Th response may switch from Th1 response in the acute phase to mixed Th1/Th2 response in the chronic phase of disease [63]. There is some speculation that NK cells may impart some protective role.

TGF-β2 provides significant therapeutic benefit [64]. The lack of trefoil factor and TGF-β1 aggravates disease production [65, 66]. The mediators involved and the efficacy of various compounds tested in this model are listed in Table 1. While many classes of compounds showed significant efficacy at various disease levels, it is disappointing that corticosteroids are not effective in this model, suggesting a variance from human disease. However, 5-ASA and congeners of 5-ASA show marginal beneficial effects in established disease, but have lower efficacy levels in "chronic" disease.

The advantage of this model also resides in the ease of producing both "acute" and "chronic" disease by simple modification of the DSS feeding protocol. The disease is highly reproducible. Chronic inflammation in this model lasts for a longer period, permitting the evaluation of the efficacy of compounds without any inherent risk of self healing. The functional end points are easy to measure, but labor intensive. Since disease activity index shows an excellent correlation with architectural changes, it can be used to quickly screen compounds to facilitate decision-making for thorough screening by histology and mediator measurements.

Most importantly, this model is highly applicable for understanding the multistep neoplastic process involved in colitis-associated colon cancer and to test investigational drugs that are chemopreventive or therapeutic to combat colon cancer in high-risk IBD patients. The disadvantage of the model is that DSS is very expensive. Batch to batch variations in disease severity could occur due to small molecular weight DSS impurities in the DSS preparation. The disease is characterized by progressive crypt dropout, suggesting a direct effect of DSS on the epithelial cells as opposed to lamina propria cells as suggested in human IBD. The disease is patchy, crypt abscesses are infrequent, and corticosteroids are ineffective, making this model not fully relevant to human UC.

Miscellaneous chemically induced models

A low molecular weight (20–40 kDa) carrageenan, a sulfated polysaccharide derived from red seaweed, given as a 5% solution, produces colitis in guinea pigs [67]. The disease spares the small intestine, first appears in the cecum within 1–2 weeks after the administration, and gradually progresses into distal colon and rectum covering the entire colon. The disease usually disappears after discontinuation of carrageenan. The lesions are localized to the mucosa and lamina propria, with significant influx of polymorphonuclear cells, crypt dropouts, non-granulomatous crypt abscess and crypt distortion. The mechanism of injury appears to be related to the presence of bacteria, particularly *Bacteroides vulgatus* [68]. This is no longer a popular preclinical testing model.

It is commonly known that certain nonsteroidal anti-inflammatory agents such as indomethacin produce gastric and intestinal inflammation and ulceration. There is sufficient evidence to support that the mechanism of injury in this model is mediated by inhibition of prostaglandins in association with bile. Evidence for this alleged mechanism came from studies in which administration of prostaglandins and their analogs prevented formation of lesions [69]. Further studies showed that luminal bacteria and their byproducts contributed to the emergence of lesions since germ-free rats failed to develop severe lesions and antibiotics attenuated the disease in this model [69, 70]. The morphological and histological changes of chronic transmural inflammation and ulceration of the small intestine observed in this model show some resemblance to CD. It appears that the model is easy to produce in rats (mice do not develop lesions) and is reproducible. The model also responds to 5-ASA, a conventional therapy used in IBD.

The remaining chemically induced models of less preclinical importance includes peroxynitrite [71], mitomycin [72], iodoacetamide [73] and phorbol-myristate acetate (PMA) models produced in rats (Tab. 1) [74]. Peroxynitrite injury is NO mediated. The model may help our understanding of the role played by NO in IBD. The iodoacetamide and N-methylmaleimide (sulfhydryl alkylator that blocks sulfhydryl groups) model [75] is produced by instilling these compounds in the colon, which produces multifocal mucosal erosions and ulcerations that resolves within 3 weeks. The mitomycin model of colitis is another model that is relatively easier to produce. Injury in this model is produced by intraperitoneal injection of mitomycin c at 2.75 mg/kg in Fisher rats. It produces diffuse colitis affecting mainly the superficial epithelium and causing mucosal barrier disruption. The PMA-induced model of colonic inflammation in rabbits was first described by Fretland et al. in 1990 [74]. Like any other experimental colitis model, it is easy to produce by giving an enema comprising of 1.5–3 mg/kg dissolved in 10 ml 20% ethanol. Colitis in this model lasted for at least 4 days. The mechanisms of injury are probably related to up-regulation of protein kinase C. Neutrophils do not seem to be directly involved in precipitating injury in this model. The model is untested for application to preclinical trials.

Genetically engineered models

Recently several inbred animals have been developed to investigate the influence of background genes on the development of colitis. Many of these are experimental in nature for understanding pathogenesis of IBD, and they may also be useful for testing agents that regulate the mediators produced through these engineered pathways. Table 1 lists several of these models. These inbred models are now used to facilitate single gene mutations in the MHC class I polymorphic molecules such as HLA-B-27, β2-microglobulin and several cytokine genes to understand their role in regulating T cell and immune functions in various diseases. These mutated, inbred models have contributed to significant breakthroughs, and will eventually open avenues to exploit pharmacological testing of many genetic, immunological and cytokine-specific targets of inflammation (Tab. 2). Several transgenic models of colitis have been described. A few of the important models among them are described below.

HLA-B27/β2 transgenic microglobulin model

In this transgenic model, rats are genetically engineered to overexpress HLA B 27/β2-microglobulin using recombinant DNA technology [76]. These animals develop spontaneous spondyloarthropathies and inflammation of multiple organs including gastrointestinal inflammatory disease. The transgenic lines have been maintained in MHC compatible F33 and Lew backgrounds. Only those homozygous animals that bear the highest gene copies of 21-4H and 33-3 lines develop diarrhea and other clinical symptoms, such as arthritis, myocarditis, uveitis and skin inflammation, which occur at variable frequency [77]. The disease affects the stomach, small and large intestines, showing some characteristics of CD and extraintestinal manifestations. In the small intestine, it produces patchy, non-granulomatous inflammation. In the colon, the disease is diffuse with crypt abscesses and heavy influx of mononuclear cells without any evidence of remission. The onset varies between 6 and 20 weeks with a male preponderance of disease. The mechanism of injury has many features of MHC class I antigenic responses. There is clear evidence that several bacterial species are involved in the disease process in this model. Antibiotic combination vancomycin and impenem is effective in controlling inflammation.

STAT 4 transgenic mice

FVB/NHSC mice, in which the regulatory transfection factor STAT 4 (signal transducer and activation transcription) is engineered transgenically, produce high levels of STAT 4 in response to systemic administration of dinitrophenylated keyhole limpet protein in Freund's adjuvant [78]. after 1–2 weeks, these animals develop severe chronic transmural colitis associated with diarrhea and weight loss. When transmural lymphocytes are stimulated *in vivo* or *in vitro* in this model, an elevated Th1 response and release of corresponding cytokines TNF-α and IFN-γ are seen due to IL-12 signaling. The colitis in this model is transferable to SCID mice when transgenic CD4 cells activated by bacterial antigens are adaptively transferred. The model is applicable to test agents which regulate Th1 responses and interfere with IL-12 signaling.

Transgenic mice expressing dominant negative N-cadherin

Cadherins are transmembrane glycoproteins that mediate cell adhesion. They are classified into multiple subclasses and play a role in morphogenesis and normal development. Increased expression of dominant negative N-cadherin interferes with normal expression of E-cadherin which is essential for formation and maintenance of epithelia. The chimeric model created by Hermiston and Gordon [79] shows transmural inflammation, goblet cell depletion, CD-like, linear and aphthous ulcers and crypt abscesses due to disruption in epithelial barrier possibly allowing normal bacterial antigens to elicit abnormal immune response and inflammation. The model may be useful for studying agents that help maintain epithelial integrity.

The isolation of mammalian genes has been of utmost importance in our understanding of various diseases. In recent years mice deficient of targeted genes have served as an important tool for modeling IBD and evaluating drugs and bacterial interactions

IL-2 knockout mice

In 1993, Sadlack et al. [80] reported that mice homologous for a disrupted IL-2 cytokine gene (IL-2$^{-/-}$) grow normally until they are about 4–5 weeks old. However, they develop multiorgan disease including colitis when they are about 35 days old. Before they are 9 weeks old, nearly 50% of animals die of anemia. The surviving animals have a normal small intestine, but develop severe colitis, gross bleeding per rectum and rectal prolapse and die within 10–25 weeks. The disease has much resemblance to human UC since the rectum is more inflamed, and then the disease progresses proximally, forming pancolitis. Histologically, the colon shows gross ulcerations, epithelial thickening due to hyperplasia, crypt abscesses and crypt distortion. The lamina propria is filled with both acute and chronic inflammatory infiltrates.

The mechanism of injury does not seem to involve many special pathogens, but involves commensal microorganisms since these animals, when derived in a pathogen-free environment, do not develop colitis. This seems to suggest that disruption of IL-2 gene causes a defect in the immune system. The abnormal immune response is associated with an increase in IL-1, IFN-γ and TNF-α, eliciting a Th1 response and IL-12 signaling. A decrease in IL-4 and IL-10 transcripts has also been noticed, suggesting that the immune response may be shifted to Th2 response, which may ultimately lead to breakdown of self tolerance. Backcross of IL-2$^{-/-}$ with RAG-2-deficient mice suggests the involvement of T lymphocytes [81]. This is further confirmed by backcrossing IL-2$^{-/-}$ with β2-microglobulin$^{-/-}$ mice which implicates CD4$^+$ as opposed to CD8$^+$ T cells [82]. Treatment with anti-IL-12 mAb shows improvement in disease (Tab. 1). The model is now extensively used to study the role played by this cytokine in pathological immune cascades. IL-2$^{-/-}$ mice, when challenged with TNP-KLH in Freund's adjuvant, shows a lack of IL-4 and TGF-β response; however, when these mice are treated with anti-CD3 antibodies, it elicits an IL-4 and TGF-β response and prevents colitis.

IL-10 knockout mice

In 1993, Kuhn et al. [83] first produced IL-10 knockout mice that develop enterocolitis characteristic of CD. Entercolitis may have resulted from the lack of IL-10 since IL-10 regulates IL-12 and TNF-α production, and lack of IL-10 results in inadequate TGF-β response, which is critical for maintaining immune homeostasis. Interestingly, recent studies have also shown that mice with a defect in IL-10 signal transduction, such as mice deficient in CRF2-4 and mice in which neutrophils and macrophages do not express STAT 3, develop colitis similar to IL-10$^{-/-}$ mice [84–86].

IL-10$^{-/-}$ mice develop normally until they are 3 weeks old. At about 7–11 weeks of age they lose significant body weight, increased mucosal permeability, and develop anemia and segmental enterocolitis involving duodenum, jejunum and proximal colon with occasional perforated ulcers. The intestine shows a varying degree of inflammation with epithelial hyperplasia cells and crypt branching. The lamina propria is filled with a heavy influx of macrophages, neutrophils and occasional multinucleate giant cells. The mechanism of injury involves common enteric microorganisms, since animals housed in a germ-free environment develop mild colitis as opposed to enterocolitis seen in animals housed in pathogen-containing environments. Abnormal immune activation is common in this model with increased expression of MHC class II molecules, and the model also shows suppression of T cell tolerance to enteric antigens. Anti-IFN-γ and anti-IL-12 antibodies inhibits inflammation [87], suggesting a Th1 response contributing to the disease process (Tab. 2).

T cell receptor knockout mice

The functional T cell receptor consists of a heterodimeric surface receptor which encompasses an α and β chain or a γ and δ chain. Mice with a disrupted gene at the α or β locus are selectively deficient of the αβ receptor. Likewise, mice with a mutated gene at the δ locus lack the γδ receptor. T cell receptor knockout (αβ and γδ$^{-/-}$) mice were produced by creating a double mutant from crossing β mutant mice with δ mutant mice [88]. Control mice deficient in mature T and B cells were also produced by mutating the recombination activating gene (RAG-1). The TCR α and β mutant mice develop normally for 12–16 weeks, and they later, in a year, develop chronic colitis reminiscent of human UC in 30% of animals. These animals do not develop colitis under germ-free conditions. They develop diarrhea, weight loss and rectal prolapse, leading to high mortality due to wasting syndrome. Gross examination of the colon shows dilatation and thickening, mostly in the rectum, and in the most severe cases, the entire colon is affected. The small intestine is not affected. Histologically, particularly in the rectum, crypts are elongated, branched and distorted. Both acute and chronic inflammatory infiltrates are seen in the lamina propria with occasional crypt abscesses. The muscularis mucosa or submucosa is usually spared of inflammation unless the disease is very severe. The mechanism by which these animals develop colitis is emerging. It appears the mechanism involves the activation of Th2 response with increased expression of IL-4, and treatment with anti-IL-4 antibodies or cross-breeding TCR$^{-/-}$ with IL-4$^{-/-}$ results in lack or a diminished form of colitis.

It appears that there is an overproduction of IL-1α and IL-1β, but not TNF-α or TGF-β1, in the colonic mucosa, since mAbs specific to these cytokines suppressed epithelial cell proliferation and increased influx of T cells in TCRα$^{-/-}$ mice. Lymphocytes with γδ receptors are not involved in inducing the disease. The control nude, RAG-1-deficient mice, housed in the same pathogen-free environment, do not develop disease. A specific pathogen has not yet been isolated from TCR knockout animals.

Gαi2 knockout mice

The α, β and γ chains of G proteins play an important role in signal transduction processes. Mice with disrupted Gαi2 develop pancolitis when they are 8–12 weeks old [89]. They show stunted growth and gradually 75% of the animals go into wasting syndrome by the time they are 35 weeks old. The disease is very severe in the distal colon and they commonly develop rectal prolapse. Histologically, these animals develop acute and chronic inflammation with mild crypt elongation, crypt abscess, crypt distortion and gross ulcerations. During a more advanced stage of disease, fibrosis and thickening of the bowel wall become very prominent. During the course of the disease these animals also develop adenocarcinoma involving both mucosa and submucosa. The small bowel is usually normal. The mechanism appears to be related to the overproduction of the Th1 class of cytokines, particularly IL-12. The role of bacterial antigens in this model has not clearly defined; however, breeding under specific pathogen-free conditions does not prevent the disease. There is evidence to suggest that the genetic background of mice dictates the degree of disease with Gαi2$^{-/-}$ mutant on 129sv background producing the most severe disease, whereas the same mutation on 129sv and C57BL/6 crossbreed produces no disease at all.

TNF-α 3' deletion (TNF-α' knockin)

Comminelli et al. [90] showed that mice carrying an endogenous deletion of the 3'-AU-rich region of the TNF-α gene develop a new CD-like phenotype, which express disease when they are 3 weeks old, with severe inflammation developing by 8 weeks. Mice developed transmural inflammation particularly in the ileum, showing the presence of villous blunting without superficial ulceration and crypt abscess. In addition, they develop polyarthritis. They also express constitutive and inducible levels TNF-α in the tissue. This is the first experimental model to show that mutational dysregulation of TNF-α plays a role in the development of intestinal lesions similar to those seen in CD. When these animals are backcrossed with TNFRI$^{-/-}$ mice, there is no disease. However, when they are backcrossed either with TNFRII$^{-/-}$ or RAG2$^{-/-}$ mice, these animals continue to have arthritis, albeit at higher levels than animals with AU-rich deletion, suggesting that the joint inflammation seen in this model is unrelated to lymphocytes. This is a promising for preclinical studies oriented towards agents that regulate TNF levels.

Multiple drug resistant (mdr1α) gene-deficient mice

The murine gene, multi-drug resistant mdr1a, encodes a 170-kDa transmembrane protein in many tissues including intestinal epithelial cells, a subset of lymphoid cells and hematopoietic cells. The mice deficient of mdr1a expresses spontaneous colitis, which can be abrogated with antibiotic treatment, suggesting that resident bacteria plays an important role in the development of disease [91]. The disease in this animal involves transmural infiltration of T and B lymphocytes with dysregulated crypt architecture, crypt abscess and superficial ulceration resembling UC, but the immune response is similar to CD, eliciting Th1 response. However, the mechanism of disease appears to be unrelated to lymphocytic function and mdr1a deletion may cause abnormal epithelial function.

Miscellaneous transgenic and knockout mice

In addition to these well-characterized gene knockout models, other mutant cytokine knockout models such as TGF-β1 knockout [92], IL-7 transgenic [93], STAT 3 knockouts [94], mice deficient in A20 protein [95], CD40L transgenic mice [96], and keratin 8 mice have also been produced (Tab. 1) [97, 98]. The TGF-β1-deficient mice produce multiorgan inflammation, including stomach and the colon, probably due to mucosal barrier disruption caused by lack of TGF-β1. Both Th1 and Th2 responses may be involved, as cytokines from both pathways are increased. The IL-7 transgenic mice [93], which overexpress IL-7 mRNA in the colonic mucosa and IL-7 receptors on lymphocytes, develop colitis when mice are 8–12 weeks old. These animals show intestinal bleeding and the appearance of rectal prolapse. Histologically, the mucosa shows erosions, crypt abscesses, goblet cell depletion and thickening of the bowel wall. The mechanism involved in producing disease in this model remains to be elucidated, but may involve IL-7 produced from colonic epithelial cells. In the keratin-8 gene disrupted FVB/N mice [97], inflammation, hyperplasia and crypt architectural changes are seen in the cecum, colon and rectum. No homozygous mouse line has been established at this time since this gene mutation causes embryonic death in C57BL/6 and 129sv mice. STAT 3 knockout in macrophages and neutrophils results in elevated levels of TNF-α, IFN-γ, IL-1 and IL-6 when systemically challenged with LPS. These mice, when they are 20 weeks old, develop enterocoltis due to the absence of counter regulation by IL-10. Mice lacking A20 protein, which is a negative regulator of NF-κB signaling, develop severe intestinal inflammation and die after 1 week. The disease may be related to increased TNF production and lack of TLR regulation. Transgenic mice overexpressing CD40L develop multiorgan inflammation and severe colitis [96].

Adoptive transfer models

CD45RB^hiCD4^+ T cell transfer model

In 1993, Morissey et al. [98] and Powrie et al. [99] demonstrated that adoptive transfer of population from BALB/c or C.B-17 mice into C.B-17 SCID mice results in colitis and wasting syndrome beginning 3–5 weeks after transfer of cells without the physical destruction of epithelial barrier. Cotransfer of CD45RB^lowCD4^+ T or unfractionated CD4^+ T cells with CD45RB^hiCD4^+ T cells protects these animals, except when CD45RB^lowCD45 cells are derived from IL-10 knockout mice. The finding that CD45RB^lowCD4^+ cells from IL-10 knockout mice are incapable of protecting the disease suggests that the lack of IL-10-producing T regulatory cells in this pool may have contributed for those pathogenic cells to produce colitis. Cotransfer of CD4^+CD25^+ Treg (T regulatory cells) and Tr1 cells prevent colitis and wasting syndrome [100, 101] in this model, and the beneficial effect is dependent upon IL-10 and TGF-β [102]. It remains to be confirmed whether SCID and RAG2^−/− mice develop disease because of lack of CD4^+CD25^+ Tregs. There is evidence for resident bacterial flora driving colitis, as antibiotics provide beneficial effects. Taken together, the investigative mechanisms suggest that the pool of T cells contain pathogenic cells and regulatory T cells. In normal animals, regulatory T cells provide protective functions.

Histologically, these animals show an enlarged and thickened colon associated with epithelial hyperplasia, goblet cell depletion, loss of crypts, ulceration and multifocal chronic inflammation with granulomas and multinucleated giant cells involving the entire length of colon. This pathological change associates with increased MHC class II expression and increases IFN-γ levels by activating the Th1 response. The model is new and exciting, but is difficult to produce in laboratories that are not equipped with a sophisticated FACStar Plus flow cytometer. It requires very high purity (>98%; 10 to 50 000) CD45RB^hiCD4^+ T cells to be injected intravenously.

Several neutralizing anti-cytokine mAbs have been tested to understand the pathological mechanisms in this model. Identification of these mechanisms would also be ideal for drug discovery testing process. Administration of anti-IL-12-p40, anti-IFN-γ and anti-TNF-α mAbs, but not anti-IL-4, inhibits disease [104, 105]. Anti-TGF-β and IL-10R1 [103] reverse the protection afforded by CD45RB^lowCD4^+ cells [104]. Recently, Davenport et al. [106] showed that CTLA4-Ig, a soluble fusion protein that binds to B7 molecules with greater affinity than CD28, prevents colitis and skin lesions in this model. Finally, anti-CD134^+ (OX40^+) antibody has also been shown to reverse colitis [107, 108].

CD3ε26 mice

In addition to the CD45RB^hiCD4+ T model, mice that transgenically overexpress the human CD3ε26 gene (Tgε26) are athymic and are deficient in T and NK cells. When syngeneic bone marrow is infused from wild-type animals, Tgε26 mice develop colitis and wasting syndrome [109]. Infusion of NK cells and transplant of neonatal thymus before bone marrow transfer affords protection [110]. These mice also produce high levels of IFN-γ and TNF-α, showing propensity towards a Th1 response. Selective inhibition of these cytokines results in complete abrogation of inflammation in this model. These studies suggest that rats and mice with immunoregulatory defects, either by overexpression of certain cytokines and MHC molecules, or the lack of certain cytokine and signal transduction molecules, develop severe spontaneous colitis.

Concluding remarks

Experience in IBD animal modeling has shown us that an ideal animal model cannot be achieved, as the human disease is chronic, unrelenting with multiple remissions and relapses. Despite this shortcoming, the number of animal models of inflammation that relatively mimic IBD either naturally, spontaneously or phenotypically through experimental physical, immunological or genetic manipulations has significantly increased. Many of these models have helped us dissect and substantially advance our understandings of the pathogenesis of IBD, at the same time as providing an unprecedented opportunity for identifying targets for therapeutic intervention and prevention strategies. Since so many experimental animal models are now available, investigators must take into consideration the species, strains, substrains, the microenvironment in which they live and the mediators involved in each of these models, for application to preclinical testing to manipulate the immune system or to develop vaccines or gene therapy.

References

1   Price AB (1978) Overlap in the spectrum of non-specific inflammatory bowel disease "colitis indeterminate". *J Clin Pathol* 31: 567–577

2   Hay JW, Hay AR (1992) Inflammatory bowel disease: costs of illness. *J Clin Gastroenterol* 14: 309–317

3   Fiocchi C (1997) Intestinal inflammation: a complex interplay of immune and non-immune cell interactions. *Am J Physiol* 273: G769–G775

4   Plevy SE, Landers CJ, Carramanzana NM, Deem RL, Shealy D, Targan SR (1997) A role

for TNF-α and mucosal T helper 1 cytokines in the pathogenesis of Crohn's disease. *J Immunol* 159: 6276–6282

5   Fuss IJ, Neurath M, Boirivant M, Klein JS, de la Motte C, Strong SA, Fiocchi C, Strober W (1996) Disparate CD4⁺ lamina propria (LP) lymphokine secretion profiles in inflammatory bowel disease. Crohn's disease LP cells manifest increased secretion of IL-5. *J Immunol* 157: 1261–1270

6   Monteleone G, Biancone L, Marasco R, Morrone G, Marasco O, Luzza F, Pallone F (1997) Interleukin 12 is expressed and actively released by Crohn's disease intestinal lamina propria mononuclear cells. *Gastroenterology* 112: 1169–1178

7   Sandborn W (2004) Advances in biologic therapy for IBD: An expert interview with William Sandborn. Medscape Gastroenterology. www.medscape.com/viewarticle/477684

8   Pfeiffer CJ (ed.) (1985) *Animal models of intestinal disease.* CRC Press, Boca Raton

9   Sundberg JP, Elson CO, Bedigian CD, Berkenmeir EH (1994) Spontaneous heritable colitis in a new substrain of C3H/HeJmice. *Gastroenterology* 107: 1726–1735

10 Fort MM, Mozaffarian A, Stover AG, Corrreia J, Ulevitch RJ, Persing DH, Bielfeldt-Ohmana H, Probst P, Jeffery E, Fling SP, Hershberg RM (2005) A synthetic TLR4 antagonist has an antiinflammatory effects in two murine models of inflammatory bowel disease. *J Immunol* 174: 6416–6423

11 Madara JL, Podolsky DK, King NW, Sehgal PK, Moore R, Winter HS (1985) Characterization of spontaneous colitis in cotton top tamarin (*Sanguinas oedipus*) and its response to sulfasalzine. *Gastroenterology* 88: 13–19

12 Chalifoux, LV, Bronson RT (1981) Colonic adenocarcinoma associated with chronic ulcerative colitis in cotton topped marmosets (*Sanguinas oedipus*). *Gastroenterology* 80: 942–946

13 Das KM, Squillante L, Henke M, Clapp N (1990) The presence of circulating antibodies in cotton top tamarins (CTT) with spontaneous colitis against an epitope on MR 40,000 protein shared by human and CTT colon epithelial cells. *Gastroenterology* 98: A468

14 Stonerook MJ, Weiss HS, Rodriguez JV, Hernandez JI, Peck PC, Wood JD (1994) Temperature-metabolism relations in the cotton-top tamarin (*Sanguinas oedipus*) model for ulcerative colitis. *J Med Primatol* 23: 16–22

15 Watkins DI, Hodi FS, Levin NL (1988) A primate species with limited major histocompatibility complex class I polymorphism. *Proc Natl Acad Sci USA* 85: 7714

16 Clapp N, Henke M, Hansard R, Carson R, Walsh R, Widomski D, Anglin C, Fretland D (1993) Inflammatory mediator changes in cotton top tamarins (CTT) after SC-41930 anti-colitic therapy. *Agents Actions* 39: C8–10

17 Kosiewicz MM, Krishnan A, Shah M, Bentz M, Matusmoto S, Cominelli F (1998) Characterization of a new spontaneous murine model of inflammatory bowel disease. *Gastroenterology* 114: G4143

18 Strober W, Nakamura K, Kitani A (2001) The SAMP1/Yit mouse: another step closer to modeling human inflammatory bowel disease. *J Clin Invest* 107: 667–670

19   Mizoguchi A, Mizoguchi E, Bhan AK (1999) The critical role of interleukin 4 but not interferon gamma in the pathogenesis of colitis in T-cell receptors alpha mutant mice. *Gastroenterology* 116: G3637

20   Strober W, Fuss IJ, Blumberg R (2002) The immunology of mucosal models of inflammation. *Annu Rev Immunol* 20: 495–549

21   Hodgson HJF, Potter BJ, Skinner J, Jewell DP (1978) Immune complex mediated colitis in rabbits. *Gut* 19: 225–232

22   Mee AS, McLaughlin JE, Hodgson HGF, Jewell JP (1979) Chronic immune colitis in rabbits. *Gut* 20: 1–5

23   Axelsson LG, Ahlstedt S (1990) Characteristics of immune-complex induced chronic experimental colitis in rats with a therapeutic effect of sulfasalazine. *Scan J Gastroenterol* 25: 203–209

24   Cominelli F, Nast CC, Clark BD, Schindler R, Lierena R, Eysselein VE, Thompson RC (1990) Interleukin 1 (IL-1) gene expression, synthesis, and effect of specific IL-1 receptor blockade in rabbit immune complex colitis. *J Clin Invest* 86: 972–980

25   Cominelli F, Nast CC, Llerena R, Dinarello CA, Zipser RD (1990) Interleukin I suppresses inflammation in rabbit colitis. Mediation by endogenous prostaglandins. *J Clin Invest* 85: 582–586

26   Ward JM, Anver MR, Haines DC, Melhorn JM, Gorelick P, Yan L, Fox JG (1996) Inflammatory large bowel disease in immunodeficient mice naturally infected with *Helicobacter hepaticus*. *Lab Anim Sci* 46: 15–20

27   Quinn TC, Goodell SE, Mkrtichian EE, Schuffler MD, Wang SP, Stamm WE, Holmes KK (1981) *Chlamydia trachomatis* proctitis. *N Engl J Med* 305: 195–200

28   Sartor RB, Cromartie WJ, Powell DW, Schwab JH (1985) Granulomatous enterocolitis induced in rats by purified bacterial cell wall fragments. *Gastroenterology* 89: 587–595

29   McCall RD, Haskill S, Zimmermann EM, Lund PK, Thompson RC, Sartor RB (1994) Tissue interleukin 1 and interleukin-1 receptor antagonist expression in enterocolitis in resistant and susceptible rats. *Gastroenterology* 106: 960–972

30   Herfarth HH, Mohanty SP, Rath HC, Tonkonogy S, Sartor RB (1996) Interleukin 10 suppresses experimental chronic, granulomatous inflammation induced by bacterial cell wall polymers. *Gut* 39: 836–845

31   Yamada T, Zimmerman T, Specian RD, Grisham MB (1993) Chronic granulomatous colitis induced by intramural injection of Freund's complete adjuvant. *Gastroenterology* 104: A804

32   MacPherson BR, Pfeiffer CJ (1978) Experimental production if diffuse colitis in rats. *Digestion* 17: 135–150

33   Strober W (1985) Animal models of inflammatory bowel disease – an overview. *Dig Dis Sci* 30: 3S–10S

34   Morris GP, Beck PL, Herridge MS, Depew WT, Szewczuk MR, Wallace JL (1989) Hapten-induced model of chronic inflammation and ulceration in the rat colon. *Gastroenterology* 96: 795–803

35   Neurath MF, Fuss I, Kelsall BL, Stuber F, Strober W (1995) Antibodies to interleukin-12 abrogate established experimental colitis in mice. *J Exp Med* 182: 1281–1290

36   Miller MJS, Sadowka-Kowicka H, Chotinnaueml S, Kakkis JL, Clark DA (1993) Amelioration of chronic ileitis by nitric oxide synthesis inhibition. *J Pharmacol Exp Ther* 264: 11–16

37   Shibata Y, Taruishi M, Ashida T (1993) Experimental ileitis in dogs and colitis in rats with trinitrobenzenesulfonic acid-colonoscopic and histopathologic changes. *Gastroenterol Japonica* 28: 518–527

38   Goldhill JM, Burakoff R, Donovan V, Rose K, Percy WH (1993) Defective modulation of colonic secretomotor neurons in a rabbit model of colitis. *Am J Physiol* 264: G671–G677

39   Elson CO, Beagley KW, Sharmanov AT, Fujihashi K, Kiyono H, Tennyson GS, Cong Y, Black, CA, Ridwan BW, McGhee JR (1996) Hapten-induced model of murine inflammatory bowel disease: mucosa immune responses and protection by tolerance. *J Immunol* 157: 2174–2185

40   Cavini A, Hackett CJ, Wilson KJ, Rothbard JB, Katz SI (1995) Characterization of epitopes recognized by hapten-specific CD4$^+$ T cells. *J Immunol* 154: 1232–1238

41   Ekstrom GM (1998) Oxazolone-induced colitis in rats: effects of budesonide, cyclosporine A, and 5-aminosalicylic acid. *Scand J Gastroenterol* 33: 174–179

42   Boirivant M, Fuss IJ, Chu A, Strober W (1998) Oxolozone colitis: a murine model of T helper cell type 2 colitis treatable with antibodies to interleukin 4. *J Exp Med* 188: 1929–1939

43   Ohkusa T (1985) Production of experimental ulcerative colitis in hamsters by dextran sulfate sodium and a change of intestinal microflora. *Jpn J Gastroenterol* 82: 1327–1336

44   Okayasu I, Hatekeyama S, Yamada M, Ohkusa T, Inagaki Y, Nakaya R (1990) A novel method in the induction of reliable experimental acute and ulcerative colitis in mice. *Gastroenterology* 98: 694–702

45   Cooper HS, Murthy SNS, Shah RS, Seergran DJ (1993) Clinicopathologic study of dextran sulfate sodium experimental murine colitis. *Lab Invest* 69: 238–249

46   Murthy SN, Cooper HS, Shin H, Shah RS, Ibrahim SA, Sedergran DJ (1993) Treatment of dextran sulfate sodium-induced murine colitis by intracolonic cyclosporin. *Dig Dis Sci* 38: 1722–1734

47   Okayasu I, Ohkusa T, Kajiuar K, Kanno J, Sakamoo S (1996) Promotion of colorectal neoplasia in experimental colitis. *Gut* 39: 87–92

48   Iwanaga T, Hoshi O, Han H, Fujita T (1994) Morphological analysis of acute ulcerative colitis experimentally induced by dextran sulfate sodium in the guinea pig: possible mechanisms of cecal ulceration. *J Gastroenterol* 29: 430–438

49   Domek MJ, Iwata F, Blackman EI, Kao J, Baker M, Vidrich A, Leung FW (1995) Antineutrophil serum attenuates dextran sulfate sodium-induced colonic damage in the rat. *Scand J Gastroenterol* 30: 1089–1094

50   Yamada T, Ohkudas T, Okayasu I (1992) Occurrence of dysplasia and adenocarcinoma

after experimental chronic ulcerative colitis in hamsters induced by dextran sulfate sodium. *Gut* 33: 1521–1527

51   Tamaru K, Kobayashi H, Koshimoto S, Kajiyama G, Shimamoto F, Brown WR (1993) Histochemical study of colonic cancer in experimental colitis in rats. *Dig Dis Sci* 38: 529–537

52   Cooper HS, Everley L, Chang W, Pfeiffer G, Lee B, Murthy S, Clapper M (2001) The role of mutant APC in the development of dysplasia and cancer in the mouse model of dextran sulfate sodium-induced colitis. *Gastroenterology* 121: 1407–1416

53   Roediger WE, Moore J, Babdige W (1997) Colonic sulfide in pathogenesis and treatment of ulcerative colitis. *Dig Dis Sci* 42: 1571–1579

54   Bylund-Fellenius AC, Landstrom E, Axelsson LG, Midtvedt T (1994) Experimental colitis induced by dextran sulfate in normal and germfree mice. *Microb Ecol Health Dis* 7: 207–215

55   Rath HC, Schultz M, Freitag R, Dieleman LA, Li F, Hans-Jörg Linde H, Schölmerich J, Sartor RB (2001) Different subsets of bacteria induce and perpetuate experimental colitis in rats and mice. *Infect Immun* 69: 2277–2285

56   Rachmilewitz D, Katakura K, Karmeli F, Hayashi T, Reinus C, Rudensky B, Akira S, Takeda K, Lee J, Takabayashi K, Raz E (2004) Toll-like receptor 9 signalling mediates the anti-inflammatory effects of probiotics in murine experimental colitis. *Gastroenterology* 126: 520–528

57   Verdu EF, Bercik P, Cukrowska B, Farre-Castany MA, Bouzourene H, Saraga E, Blum AL, Corthesy-Theulaz I, Tlaskalova-Hogenova H, Michetti P (2000) Oral administration of antigens from intestinal flora anaerobic bacteria reduces the severity of experimental colitis in BALB/c mice. *Clin Exp Immunol* 12: 46–50

58   Setoyama H, Imaoka A, Ishikawa H, Umesaki Y (2003) Prevention of gut inflammation by *Bifidobacterium* in dextran sulfate-treated gnotobiotic mice associated with *Bacteroides* strains isolated from ulcerative colitis patients. *Microbes Infect* 5115–5127

59   Mähler M, Bristol IJ, Leiter EH, Workman AE, Birkenmeier EH, Charles O, Elson CO, Sundberg JP (1996) Differential susceptibility of inbred mouse strains to dextran sulfate sodium-induced colitis. *Am J Physiol* 274: G544–G551

60   Dielman LA, Ridwan BU, Tennyson GS, Beakley KW, Elson CO (1994) Dextran sulfate sodium (DSS)-induced colitis occurs in severe combined immunodeficient (SCID) mice. *Gastroenterology* 107: 1722–1734

61   Axelsson LG, Landstrom E, Goldschmidt TJ, Gronberg A, Bylund-Fellinius A-C (1996) Dextran sulfate sodium (DSS) induced experimental colitis in immunodeficient mice. Effects in CD4[+]-cell depleted, athymic and NK-cell depleted SCID mice. *Inflamm Res* 45: 181–191

62   Egger B, Bajaj-Elliott M, MacDonald TT, Inglin R, Eyesellin VE, Buchler MW (2000) Characterization of acute murine dextran sulphate colitis: cytokine profile and dependency. *Digestion* 62: 240–248

63   Dielemann LA, Palmen MJ, Akol H, Bloemena E, Pena AS, Meuwissen SG, Van Rees EP

(1998) Chronic experimental colitis induced by dextran sulphate sodium (DSS) is characterized by Th1 and Th2 cytokines. *Clin Exp Immunol* 114: 385–391

64 Murthy SN, Cooper HS, Coppola D, Barrish S, McKibbin R, Cerletti N, DiMuzzio J (1992) Transforming growth factor b2, but not epidermal growth factor yields protection against dextran sulfate-mediated colitis in mice. *Gastroenterology* 102: A669

65 Mashimo H, Wu DC, Podolsky DK, Fishman MC (1996) Impaired defense of intestinal mucosa in mice lacking intestinal trefoil factor. *Science* 274: 204

66 Egaer B, Procaccino F, Laksmanan J, Reinshagen M, Hoffman P, Patel A, Reuben W, Gnanakkan S, Liu L, Barajas L, Eyesselein VE (1997) Mice lacking transforming growth factor alpha have an increased susceptibility to dextran sulfate-induced colitis. *Gastroenterology* 113: 825–832

67 Onderdonk AB (1985) The carrageenan model of experimental ulcerative colitis. *Prog Clin Biol Res* 186: 237–245

68 Onderdonk AB, Bronson R, Cisneros R (1987) Comparison of *Bacteroides vulgatus* strains in the enhancement of experimental ulcerative colitis. *Infect Immun* 55: 835–836

69 Yamada T, Deitch E, Specian RD, Perry MA, Sartor RB, Grisham MB (1993) Mechanisms of acute and chronic intestinal inflammation induced by indomethacin. *Inflammation* 17: 641–662

70 Banerjee AK, Peters TJ (1990) Experimental non-steroidal anti-inflammatory drug-induced enteropathy in the rat: similarities to inflammatory bowel disease and effect of thromboxane synthesis inhibitors. *Gut* 31: 1358–1364

71 Rachmilewicz D, Stanler JS, Karmeli F, Mullins ME, Singel DJ, Loxcalzo J, Xavier RJ, Podolsky DK (1993) Peroxynitrite-induced rat colitis. A new model of colonic inflammation. *Gastroenterology* 105: 1681–1688

72 Keshavarzian A (1992) Mitomycin C-induced colitis in rats: A new model of acute colonic inflammation implicating reactive oxygen species. *J Lab Clin Med* 120: 778–791

73 Rachmilewicz D, Karmeli F, Okon E (1995) Sulfhydryl blocker-induced rat colonic inflammation is ameliorated by inhibition of nitric oxide synthase. *Gastroenterology* 109: 98–106

74 Fretland DJ, Widomski DL, Levin S, Gaginella TS (1990) Colonic inflammation in the rabbit induced by phorbol-12-myristate-13-acetate. *Inflammation* 14: 143–150

75 Satoh H, Sato F, Takami K, Szabo S (1997) New ulcerative colitis model induced by sulfhydryl blockers in rats and the effects of antiinflammatory drugs on the colitis. *Jap J Pharmacol* 73: 299–309

76 Hammer RE, Maika SD, Richardson JA, Tang JP, Taurog JD (1990) Spontaneous inflammatory disease inflammatory disease in transgenic rats expressing HLA-B27 and human beta 2 m: an animal model of HLA-B27 associated disorders. *Cell* 63: 1099–1112

77 Taurog JD, Maika SD, Simmons WA, Breban M, Hammer RE (1993) Susceptibility to inflammatory disease in HLA-B27 transgenic rat lines correlates with the level of B27 expression. *J Immunol* 150: 4168–4178

78 Wirtz S, Finotto S, Kanzler S, Lohse AW, Blessing M, Leher HA, Galle PR, Neurath MF

(1999) Chronic intestinal inflammation in STAT-4 transgenic mice: characterization of disease and adoptive transfer by TNF- plus IFN-gamma-producing CD4+ T cells that respond to bacterial antigens. *J Immunol* 162: 1884–1888

79 Hermiston ML, Gordon JI (1995) Inflammatory bowel disease and adenomas in mice expressing a dominant negative N-cadherin. *Science* 270: 1203–1207

80 Sadlack B, Merz H, Schorle H, Schimpl A, Feller AC, Horak I (1993) Ulcerative colitis-like disease in mice with a disrupted interleukin-2 gene. *Cell* 75: 253–261

81 Ma A, Datta M, Margosian E, Chen J, Horak I (1995) T cells, but not B cells, are required for bowel inflammation in interleukin deficient mice. *J Exp Med* 182: 1567–1572

82 Simpson SJ, Mizoguchi E, Allen D, Bhan AK, Terhorst C (1995) Evidence that CD4+ but not CD8+ T cells are responsible for bowel inflammation in interleukin-2 deficient colitis. *Eur J Immunol* 25: 2618–2625

83 Kuhn R, Lohler J, Rennick D, Rajewsky K, Muller W (1993) Interleukin-10 deficient mice develop chronic enterocolitis. *Cell* 75: 263–274

84 Suzuki A, Hanada T, Mitsuyama K, Takafumi Y, Kamizano S, Hoshino T, Kubo M, Yamashita A, Okabe M, Takeda K et al (2001) CIS3/SOCS3/SSI3 plays a negative role in Stat 3 activation and intestinal inflammation. *J Exp Med* 193: 471–478

85 Shull MM, Ormsby I, Kier AB, Pawloski S, Diebod RJ, Yin M, Allen R, Sidman C, Proetzel G, Calvin D et al (1992) Targeted disruption of the mouse transforming growth factor-b1 gene results in multifactorial inflammatory bowel disease. *Nature* 359: 693–699

86 Spencer SD, Di MF, Hooley J, Pitts MS, Bauer M, Ruyan AM, Sordat B, Gibbs VC, Aguet M (1998) The orphan receptor CRF2-4 is an essential subunit of the interleukin 10 receptor. *J Exp Med* 187: 571–578

87 Berg DJ, Davidson N, Kuhn R, Muller W, Menon S, Holland G, Thompson-Snipes L, Leach MW, Rennick D (1996) Enterocolitis and colon cancer in interleukin-10-deficient mice are associated with aberrant cytokine production and CD4+ Th1-like responses. *J Clin Invest* 98: 1010–1020

88 Mombaeerts P, Mizoguchi E, Grusby MJ, Glimcher LH, Bhan AK, Tonegawa S (1993) Spontaneous development of inflammatory bowel disease in T cell receptor mutant mice. *Cell* 75: 275–282

89 Rudolph U, Finegold MJ, Rich SS, Harriman GR, Srinivasan Y, Brabet P, Boulay G, Bradley A, Bimbauer L (1995) Ulcerative colitis and adenocarcinoma of the colon in Gai2-deficient mice. *Nat Genet* 10: 141–148

90 Cominelli, F, Kontoyiannis D, Pizzaro TT, Kollias G (1998) Mice carrying an endogenous deletion of the 3'-AU-rich region of the TNFα gene develop a Crohn's disease-like phenotype: A key role of the TNFα in the pathogenesis of chronic intestinal inflammation. *Gastroenterology* 114: G3911

91 Panwala CM, Jones JC, Viney JL (1998) A novel model of inflammatory bowel disease: mice deficient for the multiple drug resistance gene, mdr1a, spontaneously develop colitis. *J Immunol* 161: 5733–5744

92 Diebold RJ, Eis MJ, Yin MY, Ormsby I, Boivin GP, Darrow BJ, Saffitz JE, Doetschman T (1995) Early onset of multifocal inflammation in the transforming growth factor β1-null mouse is lymphocyte mediated. *Proc Natl Acad Sci USA* 92: 12215–12219

93 Ueno Y, Watanabe M, Yamazaki M, Yajima T, Ishiii H, Uehira M, Nishimoto H, Hata J, Hibi T (1996) Interleukin-7 transgenic mice develop chronic colitis with overexpression of IL-7 mRNA in the colonic mucosa. *Gastroenterology* 110: A1033

94 Takeda K, Clausen BE, Kaisho T, Tsujimura T, Terada N, Forster I, Akira S (1999) Enhanced TH1 activity and development of chronic enterocolitis in mice devoid of Stat3 in macrophages and neutrophils. *Immunity* 10: 39–49

95 Lee EG, Boone DL, Chai S, Libby SL, Chien M, Lodolce JP, Ma A (2000) Failure regulate TNF-induced NF-kappaB and cell death responses in A20 deficient mice. *Science* 30: 1–4

96 Clegg CH, Rulffes JT, Haugen HS, Hoggatt IH, Aruffo A, Durham SK, Farr AG, Hollenbaugh D (1997) Thymus dysfunction and chronic inflammatory bowel disease in gp39 transgenic mice. *Int Immunol* 9: 1111–1122

97 Baribault H, Penner J, Iozzo RV, Wilson-Heiner M (1994) Colorectal hyperplasia and inflammation in keratin 8-deficient mice. *Genes Dev* 8: 2964–2973

98 Morissey PJ, Charrier K, Braddy S, Liggit D, Watson JD (1993) CD4+ T cells that express high levels of CD45RB induce wasting disease when transferred into congenic severe combined immune mice. *J Exp Med* 178: 237–244

99 Powrie F, Leach MW, Mauze S, Caddle LB, Coffman RL (1993) Phenotypically distinct subsets of CD4+ T cells induce or protect from chronic intestinal inflammation in C.B-17 SCID mice. *Int Immunol* 5: 1461–1471

100 Read S, Malmstrom V, Powrie F (2000) Cytotoxic T lymphocyte-associated antigen 4 plays an essential role in the function of CD25(+)CD4(+) regulatory cells that control intestinal inflammation. *J Exp Med* 192: 295–302

101 Barthlott B, Moncrieffe H, Veldhoen M, Atkins CJ, Christensen J, O'Garra A', Stockinger B (1997) A CD4+ T cell subset inhibits antigen-specific T-cell responses and prevents colitis. *Nature* 389: 737–742

102 Liu H, Hu B, Xu D, Liew FY (2003) CD4+CD25+ regulatory T cells cure murine colitis: the role of IL-10, TGF beta, and CTLA4. *J Immunol* 171: 5012–5017

103 Asseman C, Leach MW, Mauze S, Coffman RL (1999) An essential role for interleukin 10 in the function of regulatory T cells that inhibit intestinal inflammation. *J Exp Med* 190: 995–1004

104 Powrie F, Carlino J, Leach MW, Mauze S, Coffman RL (1996) A critical role of transforming growth factor-b but not interleukin-4 in the suppression of T helper type-1-mediated colitis by CD45RB^{low} CD4+ T cells. *J Exp Med* 183: 2669–2674

105 Maloy KJ, Salaun L, Cahill R, Dougan J, Saunders NJ, Powrie F (2003) CD4+CD25+ T (R) cells suppress innate immune pathology through cytokine-dependent mechanisms. *J Exp Med* 197: 111–119

106 Davenport CM, McAdams HA, Kou J, Mascioli K, Eichman C, Healy L, Peterson J, Murphy S, Coppola D, Truneh A (2002) Inhibition of pro-inflammatory cytokine gen-

eration by CTLA4-Ig in the skin and colon of mice adoptively transplanted with CD45RB[hi] CD4[+] T cells correlates with suppression of psoriasis and colitis. *Int Immunopharmacol* 2: 653–672

107 Malmstrom V, Shipton D, Singh B, Al-Shamkhani A, Puklavec MJ, Barclay AN, Powrie F (2001) CD134L expression on dendritic cells in the mesenteric lymph nodes drives colitis in T cell-restored SCID mice. *J Immunol* 166: 6972–6981

108 Higgins LM, McDonald SA, Whittle N, Crockett N, Shields JG, MacDonald TT (1999) Regulation of T cell activation *in vitro* and *in vivo* by targeting the OX40-OX40 ligand interaction: amelioration of ongoing inflammatory bowel disease with OX-40-IgG fusion protein, but not with an OX-40 ligand-IgG fusion protein. *J Immunol* 162: 486–493

109 Simpson SJ, Hollander G, Mizoguchi E, Bhan A, Terhorst C (1995) T lymphocytes in murine inflammatory bowel disease. *Clin Immunol Immunopathol* 76: S45–S46

110 Wang B, Shah SA, Simpson SJ, Allen D, Biron CA, Hollander GA, Terhorst C (1996) Protective role of natural killer cells in a mouse model of inflammatory bowel disease. *Gastroenterology* 110: A1042

111 Watkins DI, Hodi FS, Levin NL (1988) A primate species with limited major histocompatibility complex class I polymorphism. *Proc Natl Acad Sci USA* 85: 7714

112 Sterberg PE, Winsor-Hines D, Briskin MJ, Soleer-Ferran D, Merrill C, Mckay CR, Newman W, Ringer DJ (1996) Rapid resolution of chronic colitis in the cotton top tamarin with an antibody to a gut homing integrin alpha 4 beta 7. *Gastroenterology* 111: 1373–1380

113 Watkins PE, Warren BF, Stephens S, Ward P, Foulkes R (1997) Treatment of ulcerative colitis in the cotton top tamarin using antibody to tumor necrosis factor alpha. *Gut* 40: 628–633

114 Kosiewicz MM, Nast CC, Krishnan A, Rivera-Nieves J, Moskaluk CA, Matsumoto S, Kozaiwa K, Cominelli F (2001) Th1-type responses mediates spontaneous ileitis in a novel murine model of Crohn's disease. *J Clin Invest* 107: 695–702

115 Burns RC, Rivera-Nieves J, Moskaluk CA, Matsumoto S, Cominelli F, Ley K (2001) Antibody blockade of ICAM-1 and VCAM-1 ameliorates inflammation in the SAMP-1/Yit adoptive transfer model of Crohn's disease in mice. *Gastroenterology* 121: 1428–1436

116 Inoue T, Tsuzuki T, Matsuzaki K, Matsunaga H, Miyazaki J, Hokari R, Okada Y, Kawaguchi A, Nagao S, Itoh K et al (2005) Blockade of PSGL-1 attenuates CD14[+] monocyte cell recruitment in intestinal mucosa and ameliorates ileitis in SAMP1/Yit mice. *J Leukoc Biol* 77: 287–295

117 Rivera-Nieves J, Olson T, Bamias G, Bruce A, Solga M, Knight RF, Hoang S, Cominelli F, Ley K (2005) L-selectin, alpha 4 beta 1, and alpha 4 beta 7 integrins participate in CD4[+] T cell recruitment to chronically inflamed small intestine. *J Immunol* 174: 2343–2352

118 Axelsson LG, Ahlstedt S (1990) Characteristics of immune-complex-induced chronic

experimental colitis in rats with a therapeutic effect of sulphasalzine. *Scand J Gastroenterol* 25: 203–209

119 Cassini-Raggi V, Herbert C, Monsacchi L, Cominelli F (1994) A specific monoclonal antibody (MoAb) against interleukin-8 (IL-8) suppress inflammation in rabbit immune colitis. *Gastroenterology* 106: A661

120 Meenan J, Hommes DW, Mevissen M, Dijkhuizen S, Soule H, Moyle M, Buller HR, ten Kate FW, Tytgat GN, van Deventer SJ (1996) Attenuation of the inflammatory response in an animal colitis model by neutrophil inhibitory factor, a novel beta-2 integrin antagonist. *Scand J Gastroenterol* 31: 786–791

121 Hommes DW, Meenan J, Dijkhuizen S, ten Kate FJ, Tytgat GN, van Deventer SJ (1996) Efficacy of recombinant granulocyte colony-stimulating-factor (rhG-CSF) in experimental colitis. *Clin Exp Immunol* 106: 529–533

122 Podolsky DK, Lobb R, King N, Benjamin CD, Pepinsky B, Sehgal P, deBeaumont M (1993) Attenuation of colitis in the cotton top tamarin by anti-alpha-4 integrin monoclonal antibody. *J Clin Invest* 92: 372–380

123 Neurath MF, Pettersson S, Meyer zum Buschenfelde KH, Strober W (1996) Local administration of antisense phosphorothioate oligonucleotide to the p65 subunit of NF-kappa B abrogates established experimental colitis in mice. *Nat Med* 2: 998–1004

124 Neurath MF, Fuss I, Pasparakis M, Alexopoulou L, Haralambous S, Meyer zum Buschenfelde KH, Strober W, Kollias G (1997) Predominant pathogenic role of tumor necrosis factor in experimental colitis in mice. *Eur J Immunol* 27: 1743–1750

125 Bertran X, Mane J, Fernandez-Banares F, Castella E, Bartoli R, Ojanguren I, Esteve M, Gassull MA (1996) Intracolonic administration of zileuton, a selective 5-lipoxygenase inhibitor, accelerates healing in a rat model of chronic colitis. *Gut* 38: 899–904

126 Wallace JM, Keenan CM (1990) An orally active inhibitor of leukotriene synthesis accelerates healing in a rat model of colitis. *Am J Physiol* 258: G527–534

127 Vilaseca J, Salas A, Guarner F, Rodriguez R, Martinez M, Malagelada J (1990) Dietary fish oil reduces progression of chronic inflammatory lesions in a rat model of granulomatous colitis. *Gut* 31: 539–544

128 Wallace JL (1988) Release of platelet activating factor (PAF) and accelerated healing induced by a PAF antagonist in an animal model of chronic colitis. *Can J Physiol Pharmacol* 66: 422

129 Luck MS, Bass P (1993) Effect of epidermal growth factor on experimental colitis in the rat. *J Pharmacol Exp Ther* 264: 984–990

130 Zeeh JM, Procaccino F, Hoffman P, Aukerman SL, McRoberts JA, Soltani S, Pierce GF, Lakshamnan J, Lacey D, Eysseilein VE (1996) Keratinocyte growth factor ameliorates mucosal injury in an experimental model of colitis in rats. *Gastroenterology* 110: 1077–1083

131 Rachmilewicz D, Karmeli F, Okon E, Bursztyn M (1995) Experimental colitis is ameliorated by inhibition of nitric oxide synthase activity. *Gut* 37: 247–255

132 Duchmann R, Schmitt E, Knolle P, Meyer-zum Buschenfelde KH, Neurath M (1996) Tolerance towards resident intestinal flora in mice is abrogated in experimental colitis

and restored by treatment with interleukin-10 or antibodies to interleukin-12. Eur *J Immunol* 26: 934–938

133 Qui BS, Pfeiffer CJ, Jeith JCJ (1996) Protection by recombinant human inerleukin-11 against experimental TNB-induced colitis in rats. *Dig Dis Sci* 41: 1625–1630

134 McCafferty DM, Rioux KJ, Wallace JL (1992) Granulocyte infiltration in experimental colitis in the rat is interleukin-1 dependent and leukotriene independent. *Eicosanoids* 5: 121–125

135 Fuss IJ, Marth T, Neurath MF, Pearlstein GR, Jain A, Strober W (1999) Antiinterleukin 12 treatment regulates apoptosis of Th1 T cells in experimental colitis in mice. *Gastroenterology* 117: 1078–1088

136 Atreya R, Mudter J, Finnotto S, Mullberg J, Jostock T, Wurz S, Schuetz M, Bartsch B, Holtman MBC, Czaja F et al (2000) Blockade of IL-6 trans-signalling abrogates established experimental colitis in mice by suppression of T cell resistance against apoptosis in chronic intestinal inflammation: Evidence in Crohn's disease and experimental colitis *in vivo. Nat Med* 6: 583–588

137 Cooper HS, Murthy SNS, Shah RS, Sedergran DJ (1993) Clinicopathologic study of dextran sulfate sodium experimental colitis. *Lab Invest* 69: 238–249

138 Murthy SN, Cooper HS, Shin HS, Shah RD, Ibrahim SA, Sedergran DJ (1993) Treatment of dextran sulfate sodium-induced murine colitis by intracolonic cyclosporin. *Dig Dis Sci* 38: 1722–1734

139 Murthy S, Cooper HS, Coppola D, Shirer R (1992) Interleukin receptor-1 receptor antagonist is effective against dextran sulfate (DSS)-mediated colitis in mice. *Gastroenterology* 102: A669

140 Kojouharoff G, Hans W, Obermeir F, Mannel DN, Andus T, Scholmerich J, Gross V, Falk W (1997) Neutralization of tumor necrotic factor (TNF) but not IL-1 reduces inflammation in chronic dextran sulfate sodium-induced colitis in mice. *Clin Exp Immunol* 107: 353–358

141 Murthy SNS, Fondacaro JD, Murthy NS, Cooper HS, Bolkenius F (1994) Beneficial effect of MDL 73404 in dextran sulfate mediate colitis. *Agents Actions* 41 (special conference): C233234

142 Murthy S, Murthy NS, Coppola D, Wood DL (1997) The efficacy of BAY y 1015 in dextran sulfate model of mouse colitis. *Inflamm Res* 46: 224–233

143 Tomoyose M, Mitsuyama K, Ishida H, Toyonaga A, Tanikawa K (1998) Role of interleukin-10 in a murine model of dextran sulfate sodium-induced colitis. *Scand J Gastroenterol* 33: 435–440

144 Hamamoto N, Maermura K, Hirata I, Murano M, Sansaki S, Katsu K (1999) Inhibition of dextran sulphate (DSS)-induced colitis in mice by intracolonically administered antibodies against adhesion molecules (endothelial leukocyte adhesion molecule-1 (ELAM-1) or intracellular adhesionmolecule-1 (ICAM-1). *Clin Exp Immunol* 117: 462–468

145 Bennet CF, Kornburst D, Henry S, Stecker K, Howard R, Cooper S, Dutson S, Hall W, Jacoby HI (1997) An ICAM-1 antisense oligonucleotide prevents and reverses dextran sulfate sodium-induced colitis in mice. *J Pharmacol Exp Ther* 280: 988–1000

146 Sue CG, Wen X, Bailey ST, Jiang W, Rangwala SM, Keilbaugh SA, Flanigan A, Murthy S, Lazar MA, Wu GD (1999) A novel therapy for colitis utilizing PPAR-gamma ligands to inhibit the epithelial inflammatory response. *J Clin Invest* 104: 383–389

147 Murthy S, Cooper HS, Yoshitake H, Meyer C, Meyer CJ, Murthy NS (1999) Combination therapy of pentoxifylline and TNFα monoclonal antibody in dextran sulphate-induced mouse colitis. *Aliment Pharmacol Ther* 13: 251–260

148 Myers KJ, Murthy S, Flanigan A, Witchell DR, Butler M, Murray S, Siwkowski A, Goodfellow D, Madsen K and Baker B (2003) Antisense oligonucleotide blockade of tumor necrosis factor-α in two murine models of colitis. *J Pharmacol Exp Ther* 304: 411–424

149 Murthy S, Flanigan A, Coppola D, Buelow R (2002) RDP 58, a locally active TNF inhibitor, is effective in the dextran sulphate mouse model of chronic colitis. *Inflamm Res* 51: 522–531

150 Keshavarzian A, Morgan G, Sedghi S, Gordon JH, Doria A (1990) Role of reactive oxygen metabolites in experimental colitis. *Gut* 31: 786–790

151 Thomas TK, Will PC, Srivatsava A, Wilson CL, Harbison M, Little J (1991) Evaluation of an interleukin-1 receptor antagonist in the rat acetic acid-induced colitis model. *Agents Actions* 34: 187–190

152 Fedorak RN, Empey LR, MacArthur C, Jewell LD (1990) Misoprostol provides a colonic mucosal protective effect during acetic acid-induced colitis in rats. *Gastroenterology* 98: 615–622

153 Keshavarzian A, Maydek J, Zabihi R, Doria M, D'Astice M, Sorensen JRJ (1992) Agents capable of eliminating reactive oxygen species: Catalase, WR-2721 or Cu(II)2 (3,5-DIPS) decrease experimental colitis. *Dig Dis Sci* 37: 1866–1873

154 Onderdonk AB, Hermos JA, Dzink JL, Bartlett JG (1978) Protective effect of metronidazole in experimental ulcerative colitis. *Gastroenterology* 74: 521–526

155 Fang W, Broughton A, Jacobson ED (1977) Indomethacin-induced intestinal inflammation. *Dig Dis* 22: 749–760

156 Banarjee AK, Peeters TJ (1990) Experimental non-steroidal antiinflammatory drug-induced enteropathy in the rat: Similarities to inflammatory bowel disease and effect of thromboxane synthetase inhibitors. *Gut* 31: 1358–1364

157 Stenson WF (1986) Role of lipoxygenase products in inflammatory bowel disease. In: D Rachmilewitz (ed.): *Inflammatory bowel diseases*. Martinus Nijhoff, The Hague, 95–104

158 Rachmilewictz D, Stamler JS, Kameli F, Mullins ME, Singel DJ, Loxcalzo J, Xavier RJ, Podolsky DK (1993) Peroxynitrate-induced rat colitis. A new model of colonic inflammation. *Gastroenterology* 105: 1681–1688

159 Rachmilewitz D, Karmeli F, Okon E (1995) Sulfhydryl blocker-induced rat colonic inflammation is ameliorated by inhibition of nitric oxide synthase. *Gastroenterology* 109: 98–106

160 Fretland DJ, Widomski DL, Levin S, Gaginella TS (1990) Colonic inflammation in the rabbit induced by phorbol-12-myristate-13-acetate. *Inflammation* 14: 143–150

161 Peterson RL, Wang L, Albert L, Keith-JC J, Dorner AJ (1998) Molecular effects of recombinant human interleukin-11 in the HLA-B27 rat model of inflammatory bowel disease. *J Clin Invest* 100: 2766–2776

162 Ehrhardt RO, Ludviksson BR, Gray B, Neurath M, Strober W (1997) Induction and prevention of colonic inflammation in IL-12 deficient mice. *J Immunol* 158: 566–573

163 Morrissey PJ, Charrier K, Braddy S, Liggit D, Watson JD (1993) CD4+ T cells that express high levels of CD45RB induce wasting disease when transferred into congenic severe combine immune deficient mice. *J Exp Med* 178: 237–244

164 Ludviksson BR, Stober W, Nishikomori R, Hasan SK, Erhardt RO (1999) Administration of mAb against alpha E beta 7 prevents and ameliorates immunization-induced colitis in IL2-/- mice. *J Immunol* 162: 4975–4982

165 Mizoguchi E, Mizoguchi A, Bhan AK (1997) Role of cytokines in the early stages chronic colitis in TCR alpha-mutant mice. *Lab Invest* 76: 385–397

Preclinical models of vascular inflammation

H. Andreas Kalmes and Christopher F. Toombs

Department of Inflammation Research, Amgen Washington Inc., Seattle, Washington 98119, USA

Introduction

Injury and inflammation are two processes that are tightly interwoven in the development of atherosclerosis. Russell Ross, one of the pioneers in modern atherosclerosis research, initially formulated his hypothesis about the development of atherosclerosis as a response to injury [1]. This view of atherosclerosis paid tribute to the recognition that vascular smooth muscle cells (SMCs) are one of the major cellular constituents of atherosclerotic plaque, and that these cells can be activated by tissue injury. In the healthy artery SMCs are quiescent, contractile cells in the tunica media. Upon activation, e.g., by mechanical injury of the tunica media, SMCs switch from the contractile to the synthetic phenotype, migrate into the media and proliferate, thereby forming an intimal lesion (reviewed in [2]). Based on these early findings and hypotheses, SMCs appeared to be a likely candidate for a prime cellular target in atherosclerosis, following the thinking that fewer SMCs would result in smaller plaques. Since then this paradigm has shifted.

Stable vs vulnerable plaque (anatomic features)

In recent years it has become increasingly clear that inflammation is a major factor in atherosclerosis (reviewed in [3, 4]). Adhesion of circulating monocytes to arterial endothelium and their extravasation into the intimal space, followed by their differentiation into macrophages are among the earliest events in plaque formation (reviewed in [5]). In addition, T lymphocytes have been shown to be present in early as well as advanced plaques, although their role is less clearly defined at this time.

It has also become clear that the clinically most dangerous plaques appear to be those that are most inflamed as opposed to those lesions where the degree of stenosis is most severe. The majority of myocardial infarctions are precipitated as the consequence of a rupturing or eroding asymptomatic plaque in a coronary artery.

In Vivo Models of Inflammation, Vol. II, edited by Christopher S. Stevenson, Lisa A. Marshall and Douglas W. Morgan

Plaques prone to rupture are typically those that have a high relative number of macrophages and a low relative number of SMCs (reviewed in [4]). SMCs are mainly located in the cap region of the plaque, and their presence, as well as the presence of the collagen-rich matrix they secrete, is thought to seal the plaque from the arterial lumen. Macrophages concentrate in the center of the plaques and their death contributes to the development of what is often referred to as "necrotic gruel" (reviewed in [5]). The term vulnerable plaque was created [6], describing a plaque that has a large inflamed core and a relatively thin SMC- and collagen-rich cap; this plaque type is, therefore, also referred to as "thin-cap atheroma" [7]. These plaques are not necessarily the biggest ones, and in addition stenosis of the arterial lumen can be absent due to compensatory outward remodeling of the artery. Matrix-degrading proteases that are involved in the remodeling process may also be involved in plaque rupture (reviewed in [8–10]).

As a consequence, a current paradigm of how to treat atherosclerosis has shifted from inhibition of SMCs to attenuation of plaque inflammation, with the interim goal to generate a stable plaque with a thick cap [11–15]. The events leading to leukocyte recruitment include those that are caused by noxious agents associated with hyperlipidemia, hypertension, diabetes, cigarette smoking, and homocysteinemia, and their effects can be described as tissue injury in a broader sense [16], closing the circle to the initial response-to-injury hypothesis by Ross.

Markers of inflammation (local and systemic)

Both local and systemic markers of inflammatory events have recently become associated with adverse outcomes in patients with acute coronary syndrome, unstable angina or myocardial infarction. Table 1 summarizes a number of examples of serum markers of inflammation that are prognostic of poor clinical outcomes in patients critically ill with acute coronary syndrome. More direct evidence of the inflammatory process has been provided by studies that have documented elevated plaque temperature [17, 18], as a classic sign of inflammation, or the presence of infiltrating activated T cells [19, 20]. These data have been the most compelling link between inflammation and acute plaque destabilization, but attaining these measures often involves invasive techniques such as catheterization or they can only be done on explanted (atherectomy) or post-mortem samples.

While a systemic elevation of circulating leukocytes has been correlated with coronary heart disease in humans (reviewed in [21]), the identification of a specific circulating biomarker indicative of plaque instability is clearly more desirable. As shown in Table 1, the noted elevations in proinflammatory cytokines are those that are particularly high in the cytokine cascade (such as interleukin IL-1β and tumor necrosis factor TNF-α) or those that are more generalized indicators of inflammation such as serum amyloid protein A or C-reactive protein (CRP). However, the

Table 1 - Clinical studies that have associated markers of inflammation with vascular disease

Patients[a]	Principle finding	Ref.
UA	Higher in-hospital coronary event rate in patients with elevated IL-1Ra and IL-6 on admission	[87]
UA	Elevated IFN-γ production and expansion of CD4$^+$, CD28^{null} T cells in patients with unstable angina	[19]
Recent MI	Threefold increase in recurrent coronary events in patients with highest plasma TNF-α	[88]
UA	Elevated IL-6 correlates with CRP in unstable angina patients and is associate with a complicated clinical course	[89]
UA/AMI	Elevated CRP and SAA present in majority of patients presenting with unstable angina or acute MI	[90]

[a] *UA, unstable angina; MI, myocardial infarction; AMI, acute MI; SAA, serum amyloid A.*

perceived role of CRP as merely a marker of inflammation may become transformed by Bisoendial et al. [22], who demonstrated that a proinflammatory and prothrombotic state may be induced by the infusion of recombinant CRP in humans. These data have to be interpreted with caution, however, as concerns have been raised about a potential residual endotoxin contamination of the bacterially produced CRP used in this study [23]. Hence, while the signs of inflammation are present locally and systemically and have been associated with poor prognosis in cardiovascular disease, the specificity of those signals for impending plaque rupture remains to be determined.

While CRP has been shown to be one of a number of predictors of future adverse cardiovascular events, population testing for elevated CRP as a risk factor is not yet indicated. In a recent Scientific Statement issued by the US Centers for Disease Control and Prevention and the American Heart Association (AHA/CDC Scientific Statement), the panelists provided three evidence-based recommendations: 1) There is no need for hs-CRP screening of the entire adult population as a public health measure; (2) CRP can be an independent marker of risk and may be useful as a discretionary tool for evaluating people with moderate risk; and (3) there is not enough evidence to suggest using CRP to track the efficacy of treatment.

At present there are no clinical therapies that have incorporated our current degree of knowledge about the importance of inflammation for development and progression of atherosclerosis. Although HMG-CoA reductase inhibitors (statins) and aspirin have been recognized to have some anti-inflammatory effects, they have been developed for different primary indications. A true anti-inflammatory drug to

prevent plaque rupture is still missing at this time. To further increase our understanding not only of plaque growth but also of the processes leading to plaque vulnerability, animal models have been regarded as an indispensable tool. This review
will focus on a number of widely used rodent models of atherosclerosis and arterial injury response, as well as discuss some novel attempts to develop models of
plaque rupture.

Models of vascular inflammation

This article focuses on rodent models of induced intimal vascular inflammation in
the context of atherosclerosis and the arterial response to injury. We do not cover
models of spontaneous atherosclerosis, or of non atherosclerosis-related vascular
inflammatory diseases such as vasculitis and arteritis, which have been reviewed
elsewhere [24–28]. The first three models described below are models of the arterial response to injury that have been widely used in non-atherosclerotic as well as
atherosclerotic rodents. Performed in non-hyperlipidemic animals, these procedures produce a smooth-muscle-rich neointima that resembles those found in
human restenotic lesions after coronary stent placement. In hyperlipidemic animals, the lesions tend to be more complex and show a closer resemblance to spontaneously developed atherosclerotic plaques; they start out as accumulations of
macrophage foams cells and from there progress to plaques containing macrophage
foam cells and SMCs but also acellular clefts filled with cholesterol (Fig. 1). In each
scenario there are inflammatory components involved in lesion development,
including the presence of leukocytes and the demonstrated role of inflammatory
cytokines.

Endothelial denudation

Neointima formation in response to endothelial denudation with a balloon catheter
was first demonstrated for rat and rabbit models of arterial injury response [29, 30].
In the rat model described by Clowes et al. [30], the left carotid artery is isolated
through an incision on the ventral side of the neck alongside the midline. The external carotid artery is tied off distally, a 2-French balloon catheter is inserted into the
lumen of the common carotid artery by transverse arteriotomy, repeatedly advanced
all the way down to the aorta, inflated, and withdrawn to remove the endothelium.
The procedure loosely mimics that of angioplasty in human patients with respect to
the tool used – a balloon catheter. In angioplasty patients, however, the actual procedure is obviously different in that the catheter is inflated locally in a preexisting
stenotic area. In order to simulate this fact, attempts have been made to injure arteries repeatedly, e.g., in the recently published paper by Jahnke et al. [31], who have

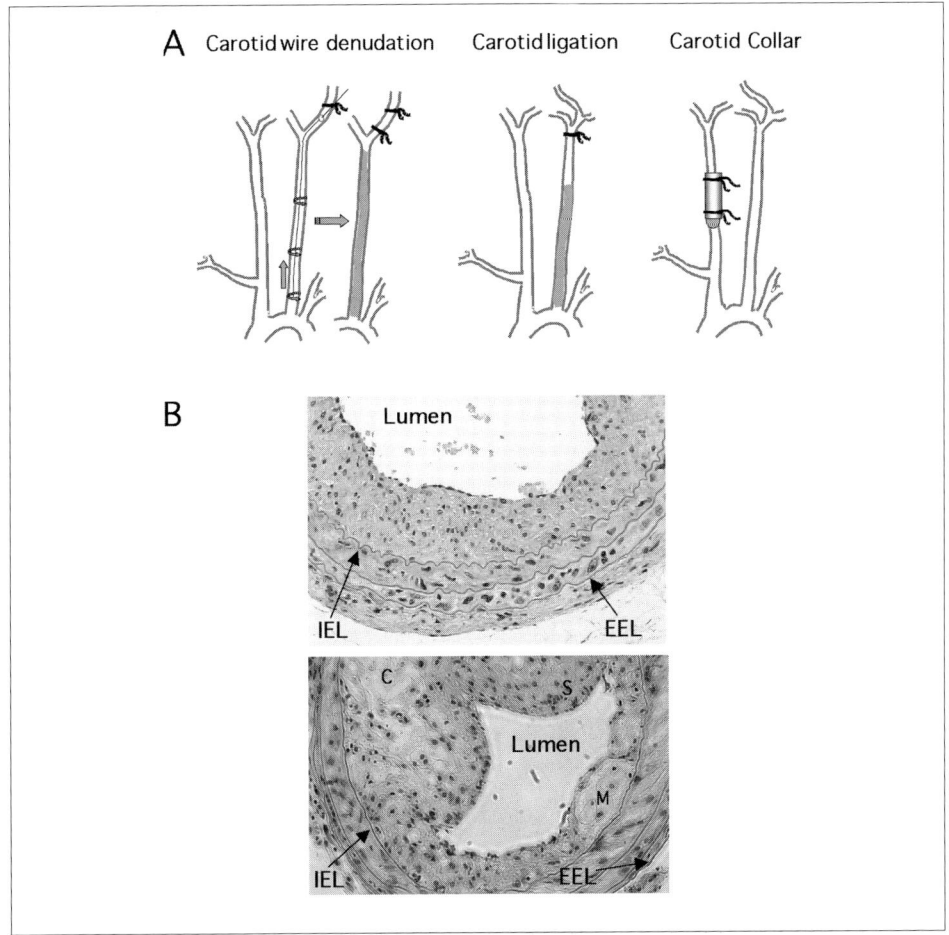

Figure 1

(A) This cartoon schematically shows the different procedures leading to intimal lesion formation in the carotid denudation model, the carotid ligation model and the carotid collar model. Lesion areas are depicted in green. (B) Carotid artery cross-sections showing typical examples of intimal lesions formed in non-hyperlipidemic (carotid ligation in a female FVB mouse, upper panel) and hyperlipidemic (carotid collar-induced plaque in a female ApoE^{−/−} mouse on atherogenic diet, lower panel) mice. Cross-sections were stained with hematoxylin and eosin. Intimal lesions are the areas between the IEL and the lumen. Whereas the neointima formed in non-hyperlipidemic mice is fairly homogeneous and mainly consists of SMCs, arterial injury in hyperlipidemic mice leads to formation of a fibrofatty plaque reminiscent of spontaneously developed atherosclerotic lesions. EEL, external elastic lamina; IEL, internal elastic lamina; C, area with cholesterol clefts; M, macrophage foam cells; S, cap area with SMCs.

combined endothelial denudation of the aorta in Lewis rats with a previously per-formed collar injury. In New Zealand White rabbits, sequential balloon injury of the aorta was used to study the processes leading to post-angioplasty restenosis [32]. The authors of this study found progressive lumen loss in the absence of any gain of the intimal area only in response to sequential, but not in response to a single, arterial injury. This form of lumen loss is similar to restenosis in response to angio-plasty in humans, which is mainly caused by elastic recoil. In contrast, the lesion that develops over 14–28 days following a single endothelial denudation in non-hyperlipidemic rodents is mainly caused by intimal hyperplasia, and thus mimics more closely the response to coronary stent placement in humans.

The principal cellular component of the neointima is the SMC, or at least a myofibroblast-like cell expressing smooth muscle alpha actin. The contribution of SMCs migrating into the intimal space from the tunica media has been demonstrat-ed (reviewed in [2]), although contributions by adventitial myofibroblasts [33–37] and more recently also by circulating precursor cells [38] have been discussed. Early papers on this subject have demonstrated that while the loss of endothelium is clear-ly important in evoking the injury response, medial damage by vessel distension is an additional contributing factor to the activation of SMCs and robust neointima formation [39–41].

Adaptation of the procedure to the mouse, developed by Lindner et al. [42], fol-lowed the desire to use knockout and transgenic strains. Due to the smaller size of mouse arteries, the mouse model does not use a balloon catheter, but instead an angioplasty guide wire that is repeatedly advanced and retracted along the common carotid artery with a rotating motion [42]. Endothelial denudation of the mouse femoral artery [43] has been used frequently as an alternative procedure, whereas a scarcely used third procedure has been described for both mouse and rat that uses a stream of dry air under pressure [44, 45]. The maximal degree of intimal injury response in mice is largely dependent on the genetic background [46]. To single out two commonly used strains for transgenic and knockout mice, respectively: FVB mice show a robust response, but the response in C57BL/6 mice is moderate at best. The same strain dependence holds true for the carotid ligation model [47].

More recently, endothelial denudation has also been used in hyperlipidemic mouse strains that are deficient for either apolipoprotein E (ApoE) or the low-den-sity lipoprotein receptor (LDLR) [48–54]. Immunohistochemical analysis revealed an increased expression of intercellular adhesion molecule-1 (ICAM-1) and vascu-lar cell adhesion molecule-1 (VCAM-1) on the luminal surface of denuded arteries in ApoE$^{-/-}$ mice [48]. Lesions formed in ApoE$^{-/-}$ mice also deficient in ICAM-1 were, however, no different than those in ICAM-1$^{+/+}$, ApoE$^{-/-}$ mice; in contrast, plaque development in the same study was suppressed in mice deficient for both ApoE and P-selectin [49].

Studies performed in this model have also confirmed the role of the chemokine CCL-2 (MCP-1, JE) and its receptor, CCR2, in plaque development: treatment with a JE/CCL2 antibody significantly inhibited monocyte arrest in ApoE$^{-/-}$ carotids 24 h after guide wire injury, and plaque development at 4 weeks post injury in mice deficient for both CCR2 and ApoE was about half of that observed in CCR2$^{+/+}$, ApoE$^{-/-}$ mice [52]. The same group reported that treatment with an antibody that neutralizes macrophage migration inhibitory factor (MIF) does not significantly reduce plaque size in response to wire injury, but leads to a more stable plaque phenotype with fewer macrophages and a higher SMC and collagen content [51]. In contrast, overall plaque size and intima/media ratio were significantly reduced by treatment with an MIF antibody in a similar study performed in LDLR$^{-/-}$ mice [45]. Lesion formation in both wild-type and ApoE$^{-/-}$ mice that underwent bilateral femoral endothelial denudation was attenuated by treatment with the non-steroidal anti-inflammatory drug (NSAID), sulindac, confirming the role of inflammation in this model [55].

The endothelial denudation model provides the opportunity to study SMC proliferation and endothelial re-growth, both of which are critical factors of restenosis in response to coronary stent placement in human patients. For an incomplete list of mouse studies see Table 2.

Flow cessation model

This model was originally developed by Kumar and Lindner and has been described in detail elsewhere [56]. Similar to the endothelial denudation model, carotid ligation with subsequent intimal lesion formation has been reported earlier for rats and rabbits, but has mainly been used in the context of stroke models. In mice, the model offers a simplified surgical procedure as compared to the endothelial wire denudation model, but results in an overall similar lesion phenotype. The common carotid artery (usually the left carotid) is isolated by blunt dissection and ligated proximal to its bifurcation with 6-0 silk suture. The ensuing lesion develops fully over a time course of 28 days, and involves remodeling and intimal hyperplasia mainly in the mid to proximal portion of the carotid artery [56]. The lesion overall seems more robust than that induced by endothelial denudation, but displays the same strain-dependent susceptibility that has been observed in the denudation model [47].

Although the lesion forming in response to flow cessation overall seems to be very similar to that seen in the denudation model, the procedure does not resemble any clinical procedure, which has been a major point of criticism. In addition, the mechanism of action for neointima formation in this model has not been clarified. It is of note, though, that intimal lesion formation in response to low blood flow (and conversely, intimal reduction after switch to higher flow), has been described in other models: Low shear stress induces intimal hyperplasia in grafted veins and in endothelialized prosthetic grafts in dogs and primates [57–60].

Table 2 - Mouse studies in the denudation, ligation, and collar models of arterial inflammation

Model	Artery	Treatment/strain	Effect	Ref.
Wire denudation	Femoral	PKCβ$^{-/-}$	↓	[91]
Wire denudation	Carotid	JE/CCL2 antibody CCR2$^{-/-}$/ApoE$^{-/-}$	↓	[52]
Wire denudation	Carotid	MIF antibody	MΦ ↓ SMCs ↑	[51]
Pressure denudation	Carotid	MIF antibody	↓	[45]
Wire denudation	Carotid	iex-1 overexpression	↓	[54]
Wire denudation	Femoral	Endostatin overexpression	↑	[92]
Wire denudation	Carotid	PAF-AH overexpression	↓	[50]
Wire denudation	Carotid	P-selectin$^{-/-}$	↓	[49]
Wire denudation	Carotid	ICAM-1$^{-/-}$	→	[49]
Wire denudation	Femoral	Sulindac	↓	[55]
Wire denudation	Femoral	Human B-Myb tg	↓	[93]
Wire denudation	Carotid	ApoE$^{-/-}$ vs. BL/6 apoE3 tg (FVB)	↑ ↓	[94]
Wire denudation	Carotid	Human Apo A-1 tg	↓	[95]
Pressure denudation	Carotid	Simvastatin	BrdU↓, leukocyte accumulation↓, apoptosis ↑	[96]
Pressure denudation	Carotid	Mac-1 (CD11b/CD18)$^{-/-}$	↓	[97]
Flow cessation	Carotid	FVIII$^{-/-}$	↓	[98]
Flow cessation	Carotid	MEKK$^{-/-}$	↓	[99]
Flow cessation	Carotid	ADAMTS transgenic	↑	[100]
Flow cessation	Carotid	IRFI-042 (lipid peroxidation inhibitor)	↓	[18]
Flow cessation	Carotid	Caveolin-1$^{-/-}$	↑	[101]
Flow cessation	Carotid	P22phox tg	↑	[102]
Flow cessation	Carotid	HS-deficient perlecan tg	↑	[103]
Flow cessation	Carotid	MMP-2$^{-/-}$	↓	[104]
Flow cessation	Carotid	TFPI(K1)$^{+/-}$	↑	[105]
Flow cessation	Carotid	TFPI tg (Ad-mTFPImyc)	↓	[106]
Flow cessation	Carotid	RIIIS/J (low plasma vWF)	↓	[107]
Flow cessation	Carotid	IκB-α tg	↓	[108]
Flow cessation	Carotid	MMP-9$^{-/-}$	↓	[109]
Wire denudation	Carotid	MMP9$^{-/-}$	↓	[110]
Flow cessation	Carotid	Endothelin receptor B$^{-/-}$	↑	[111]
Flow cessation	Carotid	P-selectin$^{-/-}$	↓	[112]
Flow cessation	Carotid	FGF-2 antibody	→ (intima), ↓ (remodeling)	[113]

Table 2 (continued)

Model	Artery	Treatment/strain	Effect	Ref.
Flow cessation	Carotid	TNF-$\alpha^{-/-}$	↓	[61]
Flow cessation	Carotid	IL-1R$^{-/-}$	↓	[61]
Collar model	Carotid	Serp-1 treatment	↓	[114]
Collar model	Carotid	VCAM-1 antibody	↓	[115]
Collar model	Carotid	p53 tg	Cap/core ratio ↓	[71]
Collar model	Femoral	p53$^{-/-}$	↑	[70]
Collar model	Carotid	IL-18 (Ad-IL-18)	Cap/core ratio ↓	[116]
Collar model	Carotid	iNOS$^{-/-}$	↓	[69]
Collar model	Carotid	IL-10 tg (Ad-IL-10)	↓	[72]
Collar model	Femoral	Ovariectomy, 17β-estradiol treatment	↑ (ovariectomy) ↓ (17β-estradiol)	[67]
Collar model	Femoral	AngII type 2 receptor (AT2)$^{-/-}$ Valsartan (AT1 inhibitor)	↑ AT2$^{-/-}$ ↓ (valsartan)	[66]
Collar model	Femoral	TLR4$^{-/-}$	↓	[68]
Collar model	Femoral	ApoE*3Leiden tg	↑	[117]
Collar model	Carotid	Human HDL i.v.	↓	[64]
Collar model	Femoral	Adrenomedullin tg	↓	[118]
Collar model	Femoral	Osteopontin tg	↑	[119]
Collar model	Femoral	Cerivastatin	↓	[120]
Collar model	Femoral	IL-1RA$^{-/-}$	↑	[121]

The neointima induced by flow cessation clearly has inflammatory components, as indicated by the presence of leukocytes: both neutrophils and macrophages have been identified [56]. Neointima formation also is significantly reduced in mice deficient for either TNF-α or the IL-1 receptor (IL-1RI) [61]. Overall, there seems to be the notion that this model has a larger inflammatory component when compared to the denudation model, but a thorough comparative analysis of both models is lacking. In an attempt to increase the physiological relevance of the carotid ligation model, two groups have recently reported a variation on the same theme, whereby flow is reduced but not completely cut off, as a consequence of ligation of some, but not all, of the arterial branches immediately distal to the carotid bifurcation [62, 63]. In this context, it is worth adding that lesion formation in the carotid endothelial denudation model might also have a flow-dependent component because the external carotid artery is permanently ligated after arteriotomy with the angioplasty guide wire.

Perivascular collar model

Following the same scheme as for the models described so far, induction of focal arterial lesions by perivascular collar placement was first described in species other than the mouse. In mice, periadventitial collar placement around either carotid or femoral artery has been used as a model both in hyperlipidemic and non-hyperlipidemic mice on a C57BL/6 background (Tab. 2). A longitudinally opened piece of plastic (silastic) tubing is slipped around the artery and closed with two or three silk sutures. Similar to the carotid ligation model, the collar model does not mimic any clinical procedure, and the mechanisms leading to plaque induction are not completely understood. Although the collar does not lead to a grossly visible obstruction of blood flow, lesion formation is likely to have a hemodynamic component.

Collar-induced neointimal lesions that develop in non-hyperlipidemic mice are similar to those seen in the previously discussed models. Although, as discussed earlier, C57BL/6 mice in the aforementioned models show at best a modest response to mechanical arterial injury [46, 47], they seem to respond with an appreciable degree of neointima formation to collar-induced injury, particularly in the femoral artery. Periadventitial collar placement in hyperlipidemic mice as a model for the induction of atherosclerotic plaque was first introduced by two different groups [64, 65]. Hyperlipidemic mice (ApoE$^{-/-}$ or LDLR$^{-/-}$ mice) are switched from normal chow to an atherogenic diet at least 1 week prior to surgery. The cellular composition and anatomy of plaques that form at the proximal end of the collar resemble those of atherosclerotic plaques that develop spontaneously in these mouse strains. They start out as intimal accumulations of foam cells and progress to complex plaques that display a core region that develops cholesterol clefts and becomes progressively acellular, as well as a collagen-rich cap that contains SMCs. Plaque formation is faster and more robust in ApoE$^{-/-}$ mice than in LDLR$^{-/-}$ mice [65].

The collar model has been used in a number of studies that have furthered or confirmed our understanding of the mechanisms involved in the arterial injury response and the development of atherosclerotic plaque (Tab. 2). In the non-hyperlipidemic "flavor", lesion development in the femoral artery has been reported to be associated with up-regulation of the chemokine MCP-1 and the cytokine TNF-α, and to be in part dependent on the activation of angiotensin receptors [66, 67]. Bacterial lipopolysaccharide (LPS) augments lesion formation in cuffed femoral arteries, and lesions are smaller in Toll-like receptor 4 (TLR4)-deficient mice when compared to wild-type mice [68]. Lesions that develop after carotid collar placement in mice deficient for inducible nitric oxide synthase (iNOS) are smaller [69], whereas those in p53-deficient mice are larger than in wild-type mice [70]. In ApoE$^{-/-}$ mice, transfection with p53 led to increased plaque cap apoptosis and a reduced cap/intima ratio; when mice were treated with phenylephrine, plaque rupture was observed [71]. In LDLR$^{-/-}$ mice, collar-induced plaques were smaller upon local overexpression of the anti-inflammatory cytokine IL-10 [72].

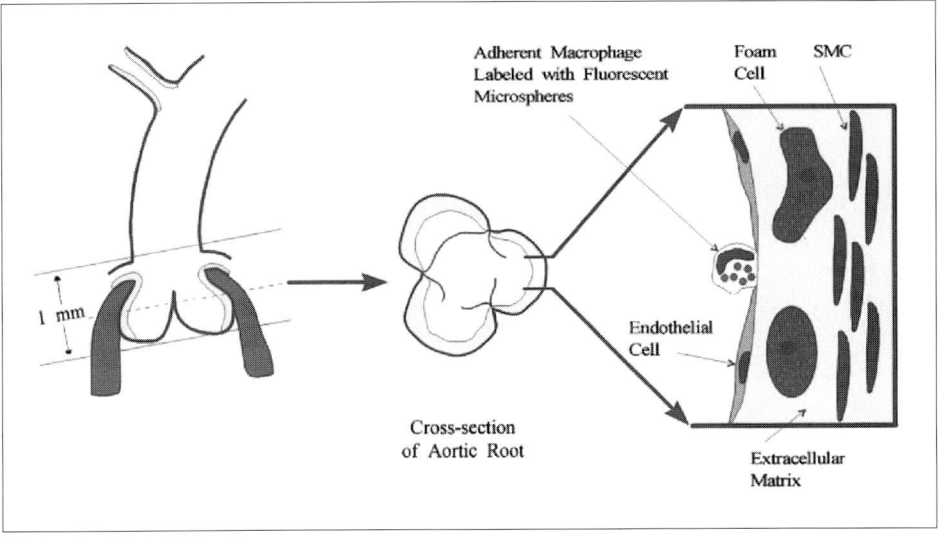

Figure 2
Schematic diagram of the study area in the aortic root, depicting a labeled macrophage adhering to the plaque.

Macrophage homing to atherosclerotic lesions *in vivo*

Patel et al. [73] have developed a model of macrophage homing to atherosclerotic lesions (Fig. 2). Activated peritoneal macrophages were labeled by coincubation with opsonized fluorescent microspheres, and then delivered intravenously to ApoE-deficient mice. After 48 h, atheromatous lesions in sections taken from the aortic root were examined by light and fluorescence microscopy to quantify the presence of fluorescently labeled macrophages. The distribution of labeled cells was highest in lung tissue and liver as well as the red pulp of the spleen. The aortic root of the ApoE-deficient mouse is the location of the most severe atheromatous lesions [74], and in these segments labeled macrophages were readily observed to be adherent or had become incorporated into the lesion. Sections from animals heterozygous in ApoE showed only minimal macrophage homing. Use of neutralizing antibodies directed against various adhesion molecules indicated that macrophage homing could be greatly attenuated with the inhibition of ICAM-l or an antibody against the alpha-subunit of the $\alpha_4\beta_1$ integrin; however, macrophage homing was unaffected when a neutralizing antibody against E-selectin was used [73]. The authors conclude the model enables studies of adhesion molecule antagonists or novel anti-inflammatory molecules in the macrophage trafficking component of atherosclerotic disease progression. A limitation of this model is the fact that in a real life scenario monocytes, and not differentiated macrophages, home to plaques.

Since the initial description of the model, slight variations on this model have been used to further elucidate mechanism of atherosclerosis. Aicher et al. [75] used fluorescence-labeled dendritic cells (DCs) in a similar fashion to show that administration of nicotine to ApoE-deficient animals that had received the labeled cells showed a threefold increase in the accumulation of DCs in the aortic root. In addition, nicotine administration led to an increased expression of costimulatory molecules and enhanced production of both pro- and anti-inflammatory cytokines (IL-12, IL-10).

Kim et al. [76] used a similar model to show recruitment of monocytes to atheroma in LDL-receptor-deficient mice using a PCR-based technique to track the Y-chromosome-specific Sry gene in the aortic roots of female LDLR$^{-/-}$ mice that had received peritoneal macrophages from male donors. In these studies, combined administration of TNF-α and IL-1β approximately doubled the accumulation of monocytes in the lesions of the aortic root. Van Lenten et al. [77] also used an adaptation of the model described by Kim et al. to demonstrate an augmented accumulation of macrophages following inoculation with influenza virus. Subsequently, it was also found that a peptide mimetic of ApoA-I was able to reduce the pro-atherogenic effects of the exposure to influenza [78].

In summary, the macrophage homing model has been used to demonstrate direct trafficking of monocytes and macrophages to atherosclerotic lesions. Known or suspected risk factors for promotion of atherosclerotic disease in humans appear to also augment monocyte-macrophage homing in this murine model. Further, the demonstration of attenuation of homing by a number of therapeutic inhibitors suggests this model is useful for assessment of candidate anti-atherosclerotic therapies.

Thermal instability of vulnerable plaque

Cascells et al. [17] and Stefanadis et al. [18] were the first to describe thermal heterogeneity in *ex vivo* human carotid endarterectomy samples and in subjects with unstable angina where the elevated plaque temperature was also strongly correlated with an elevated serum CRP. In animals, elevated plaque temperature has been documented in femoral atherosclerotic lesions in dogs and in aortas in Watanabe rabbits [79]. While much of the research in this area has focused on technology development, some limited evaluation of therapeutics has been performed. Verheye et al. [80] generated atheromatous lesions with elevated plaque temperature in New Zealand White rabbits that received an atherogenic diet for 6 months. The elevation in plaque temperature was lost upon returning the animals to normal diet for 3 months. The magnitude of the temperature gain appeared to correlate with macrophage content of the lesion and plaque thickness as determined by intravascular ultrasound.

Much of the work in this field is currently devoted to technology development around thermistors or catheters that can be used to perform thermography studies. As such, there can be reasonable hope that such technologies can be deployed in both animal and human studies to aid in the detection and diagnosis of plaque inflammation.

Vulnerable plaque

The clinical focus of atherosclerotic disease is currently on those factors that drive plaque rupture. Atherosclerosis may well be a benign disease, except for the potential of an atherosclerotic lesion to become unstable and precipitate thrombo-occlusive events. While plaque temperature was noted in the previous section as evidence of plaque inflammation, it does not necessarily signal that plaque rupture is imminent. Several hallmarks of the 'vulnerable plaque' have been identified histologically such as an enlarged lipid core, decreased collagen content or decreased overall cellularity of the cap over the lesion and presence of inflammation. As reviewed by Pasterkamp and Falk [81], such pathological features of the vulnerable plaque are very telling, yet the *ex post facto* nature of these pathology findings limits their use in the diagnosis and management of atherosclerotic disease.

Animal models of vulnerable plaque have remained elusive. While it is possible to observe evidence of intraplaque hemorrhage in ApoE$^{-/-}$ mice [82, 83], the presence of hemorrhage is, at best, evidence that a plaque rupture had occurred in the past. Unfortunately, animal models of plaque rupture (against a background of atherosclerotic disease) lack an "on demand" feature, in contrast to, for example, models of myocardial ischemia infarction, where coronary occlusions can be produced surgically.

One possible model of vulnerable plaque with an "on demand" feature was developed by Rekhter et al. [84]. The technique is surgically demanding, yet does allow for the evaluation of lesion strength, the inverse of which can potentially be considered a measure of plaque vulnerability. Briefly, balloon catheters are placed in the abdominal aorta of cholesterol-fed New Zealand White rabbits. Catheters are localized at sites of previous aortic balloon injury (to induce SMC proliferation) and the catheters are positioned such that balloons would be in constant contact with the aortic wall (to allow for SMC migration and proliferation around the balloon). After a period of 4 weeks, indwelling balloons become completely encapsulated with proliferating SMC and cholesterol-rich lesions in nearly 90% of animals studied. Aortic sections are either explanted for histology or used for plaque rupture studies *ex vivo*. The *ex vivo* plaque rupture studies allow determination of plaque vulnerability by assessing the pressure at which the inflating balloon ruptures the surrounding plaque. Inflation of balloon catheters that remain *in vivo* can be driven to the point of rupture and become the nidus for the formation of thrombus at the rupture site. Follow-on work by Rehkter et al. ([85], further reviewed in [86]) confirmed the role of hypercholesterolemia as increasing the vulnerability of such plaques to rupture.

Conclusions

As inflammation is becoming increasingly viewed as a major and independent risk factor for cardiovascular disease, longstanding models of atherosclerosis and vascular injury must now be evaluated in the context of those pathological features that develop in response to inflammation. It is quite possible that our understanding of existing therapies is about to change dramatically, when the concept of vascular inflammation is overlaid onto the progression of cardiovascular disease. Perhaps an agent as old as aspirin with its bona fide benefit in cardiovascular disease will experience a shift in the attribution of its pharmacological effects, where the anti-inflammatory effects become equally important for cardiovascular disease as the effects on platelet cyclooxygenase.

For preclinical models to be of predictive value in human disease, there must be reasonable parity of biological pathways between man and lower vertebrates. Such may be the case with elements of biology that are highly homologous across mammalian species such as norepinehrine and adrenergic receptors or heparin and anticoagulation, where strong ties have been established between preclinical and clinical pharmacology. The inflammatory process and the immune system operate in a very complex system of activating and attenuating signals, many of which are biologicals that are only partially understood or have yet to be discovered. Further, the parity between the inflammation and immune pathways between man and preclinically tested species may not be as strong as in other areas of biology. While the contribution of inflammation to vascular disease has been recognized, our ability to thoroughly understand its contribution is currently hindered, but no doubt will improve as technology improves.

References

1   Ross R, Glomset J, Harker L (1977) Response to injury and atherogenesis. *Am J Pathol* 86(3): 675–684
2   Kraiss LW, Clowes AW (1997) Response of the arterial wall to injury and intimal hyperplasia. In: A Sidawy, B Sumpio, R DePalma (eds): *The Basic Science of Vascular Disease.* Armonk, New York, 289–317
3   Ross R (1999) Atherosclerosis – an inflammatory disease. *N Engl J Med* 340(2): 115–126
4   Libby P (2001) Inflammation in atherosclerosis. *Nature* 420(6917): 868–874

5   Glass CK, Witztum JL (2001) Atherosclerosis. The road ahead. *Cell* 104(4): 503–516

6   Muller JE, Tofler GH (1992) Triggering and hourly variation of onset of arterial thrombosis. *Ann Epidemiol* 2(4): 393–405

7   Virmani R, Burke AP, Kolodgie FD, Farb A (2003) Pathology of the thin-cap fibroatheroma: a type of vulnerable plaque. *J Interv Cardiol* 16(3): 267–272

8   Shah PK (2003) Mechanisms of plaque vulnerability and rupture. *J Am Coll Cardiol* 41(4 Suppl S): 15S–22S

9   Galis ZS, Sukhova GK, Lark MW, Libby P (1994) Increased expression of matrix metalloproteinases and matrix degrading activity in vulnerable regions of human atherosclerotic plaques. *J Clin Invest* 94(6): 2493–2503

10   Carmeliet P (2000) Proteinases in cardiovascular aneurysms and rupture: targets for therapy? *J Clin Invest* 105(11): 1519–1520

11   O'Keefe JH Jr, Conn RD, Lavie CJ, Bateman TM (1996) The new paradigm for coronary artery disease: altering risk factors, atherosclerotic plaques, and clinical prognosis. *Mayo Clin Proc* 71(10): 957–965

12   Shah PK (1996) Pathophysiology of plaque rupture and the concept of plaque stabilization. *Cardiol Clin* 14(1): 17–29

13   MacIsaac AI, Thomas JD, Topol EJ (1993) Toward the quiescent coronary plaque. *J Am Coll Cardiol* 22(4): 1228–1241

14   Forrester JS (2002) Prevention of plaque rupture: a new paradigm of therapy. *Ann Intern Med* 137(10): 823–833

15   Ambrose JA, Martinez EE (2002) A new paradigm for plaque stabilization. *Circulation* 105(16): 2000–2004

16   Ross R (1995) Cell biology of atherosclerosis. *Annu Rev Physiol* 57: 791–804

17   Casscells W, Hathorn B, David M, Krabach T, Vaughn WK, McAllister HA, Bearman G, Willerson JT (1996) Thermal detection of cellular infiltrates in living atherosclerotic plaques: possible implications for plaque rupture and thrombosis. *Lancet* 347(9013): 1447–1451

18   Stefanadis C, Diamantopoulos L, Vlachopoulos C, Tsiamis E, Dernellis J, Toutouzas K, Stefanadi E, Toutouzas P (1999) Thermal heterogeneity within human atherosclerotic coronary arteries detected *in vivo*: A new method of detection by application of a special thermography catheter. *Circulation* 99(15): 1965–1971

19   Liuzzo G, Kopecky SL, Frye RL, O'Fallon WM, Maseri A, Goronzy JJ, Weyand CM (1999) Perturbation of the T-cell repertoire in patients with unstable angina. *Circulation* 100(21): 2135–2139

20   Nakajima T, Schulte S, Warrington KJ, Kopecky SL, Frye RL, Goronzy JJ, Weyand CM (2002) T-cell-mediated lysis of endothelial cells in acute coronary syndromes. *Circulation* 105(5): 570–575

21   Madjid M, Awan I, Willerson JT, Casscells SW (2004) Leukocyte count and coronary heart disease: implications for risk assessment. *J Am Coll Cardiol* 44(10): 1945–1956

22   Bisoendial RJ, Kastelein JJ, Levels JH, Zwaginga JJ, van den Bogaard B, Reitsma PH,

Meijers JC, Hartman D, Levi M, Stroes ES (2005) Activation of inflammation and coagulation after infusion of C-reactive protein in humans. *Circ Res* 96(7): 714–716

23 van den Berg CW, Taylor KE (2005) Letter in response to Bisoendial et al: "Activation of inflammation and coagulation after infusion of C-reactive protein in humans". *Circ Res* 97(1): e2

24 Jawien J, Nastalek P, Korbut R (2004) Mouse models of experimental atherosclerosis. *J Physiol Pharmacol* 55(3): 503–517

25 Yanni AE (2004) The laboratory rabbit: an animal model of atherosclerosis research. *Lab Anim* 38(3): 246–256

26 Meir KS, Leitersdorf E (2004) Atherosclerosis in the apolipoprotein-E-deficient mouse: a decade of progress. *Arterioscler Thromb Vasc Biol* 24(6): 1006–1014

27 Dal Canto AJ, Virgin HW (2000) Animal models of infection-mediated vasculitis: implications for human disease. *Int J Cardiol* 75 (Suppl 1): S37–45; discussion S47–52

28 Moyer CF, Reinisch CL (1989) Vasculitis in MRL/1 pr mice: model of cell-mediated autoimmunity. *Toxicol Pathol* 17(1 Pt 2): 122–128

29 Stemerman MB, Spaet TH, Pitlick F, Cintron J, Lejnieks I, Tiell ML (1977) Intimal healing. The pattern of reendothelialization and intimal thickening. *Am J Pathol* 87(1): 125–142

30 Clowes AW, Reidy MA, Clowes MM (1983) Mechanisms of stenosis after arterial injury. *Lab Invest* 49(2): 208–215

31 Jahnke T, Karbe U, Schafer FK, Bolte H, Heuer G, Rector L, Brossmann J, Heller M, Muller-Hulsbeck S (2005) Characterization of a new double-injury restenosis model in the rat aorta. *J Endovasc Ther* 12(3): 318–331

32 Courtman DW, Schwartz SM, Hart CE (1998) Sequential injury of the rabbit abdominal aorta induces intramural coagulation and luminal narrowing independent of intimal mass: extrinsic pathway inhibition eliminates luminal narrowing. *Circ Res* 82(9): 996–1006

33 De Leon H, Ollerenshaw JD, Griendling KK, Wilcox JN (2001) Adventitial cells do not contribute to neointimal mass after balloon angioplasty of the rat common carotid artery. *Circulation* 104(14): 1591–1593

34 Wilcox JN, Scott NA (1996) Potential role of the adventitia in arteritis and atherosclerosis. *Int J Cardiol* 54 (Suppl): S21–35

35 Wilcox JN, Cipolla GD, Martin FH, Simonet L, Dunn B, Ross CE, Scott NA (1997) Contribution of adventitial myofibroblasts to vascular remodeling and lesion formation after experimental angioplasty in pig coronary arteries. *Ann NY Acad Sci* 811: 437–447

36 Wilcox JN, Waksman R, King SB, Scott NA (1996) The role of the adventitia in the arterial response to angioplasty: the effect of intravascular radiation. *Int J Radiat Oncol Biol Phys* 36(4): 789–796

37 Scott NA, Cipolla GD, Ross CE, Dunn B, Martin FH, Simonet L, Wilcox JN (1996) Identification of a potential role for the adventitia in vascular lesion formation after balloon overstretch injury of porcine coronary arteries. *Circulation* 93(12): 2178–2187

38 Wilcox JN, Okamoto EI, Nakahara KI, Vinten-Johansen J (2001) Perivascular respons-

es after angioplasty which may contribute to postangioplasty restenosis: a role for circulating myofibroblast precursors? *Ann NY Acad Sci* 947: 68–90; discussion 90–92

39 Clowes AW, Reidy MA, Clowes MM (1983) Kinetics of cellular proliferation after arterial injury. I. Smooth muscle growth in the absence of endothelium. *Lab Invest* 49(3): 327–333

40 Clowes AW, Clowes MM, Fingerle J, Reidy MA (1989) Kinetics of cellular proliferation after arterial injury. V. Role of acute distension in the induction of smooth muscle proliferation. *Lab Invest* 60(3): 360–364

41 Fingerle J, Au YP, Clowes AW, Reidy MA (1990) Intimal lesion formation in rat carotid arteries after endothelial denudation in absence of medial injury. *Arteriosclerosis* 10(6): 1082–1087

42 Lindner V, Fingerle J, Reidy MA (1993) Mouse model of arterial injury. *Circ Res* 73(5): 792–796

43 Roque M, Fallon JT, Badimon JJ, Zhang WX, Taubman MB, Reis ED (2000) Mouse model of femoral artery denudation injury associated with the rapid accumulation of adhesion molecules on the luminal surface and recruitment of neutrophils. *Arterioscler Thromb Vasc Biol* 20(2): 335–342

44 Fishman JA, Ryan GB, Karnovsky MJ (1975) Endothelial regeneration in the rat carotid artery and the significance of endothelial denudation in the pathogenesis of myointimal thickening. *Lab Invest* 32(3): 339–351

45 Chen Z, Sakuma M, Zago AC, Zhang X, Shi C, Leng L, Mizue Y, Bucala R, Simon D (2004) Evidence for a role of macrophage migration inhibitory factor in vascular disease. *Arterioscler Thromb Vasc Biol* 24(4): 709–714

46 Kuhel DG, Zhu B, Witte DP, Hui DY (2002) Distinction in genetic determinants for injury-induced neointimal hyperplasia and diet-induced atherosclerosis in inbred mice. *Arterioscler Thromb Vasc Biol* 22(6): 955–960

47 Harmon KJ, Couper LL, Lindner V (2000) Strain-dependent vascular remodeling phenotypes in inbred mice. *Am J Pathol* 156(5): 1741–1748

48 Manka DR, Wiegman P, Din S, Sanders JM, Green SA, Gimple LW, Ragosta M, Powers ER, Ley K, Sarembock IJ (1999) Arterial injury increases expression of inflammatory adhesion molecules in the carotid arteries of apolipoprotein-E-deficient mice. *J Vasc Res* 36(5): 372–378

49 Manka D, Collins RG, Ley K, Beaudet AL, Sarembock IJ (2001) Absence of p-selectin, but not intercellular adhesion molecule-1, attenuates neointimal growth after arterial injury in apolipoprotein e-deficient mice. *Circulation* 103(7): 1000–1005

50 Quarck R, De Geest B, Stengel D, Mertens A, Lox M, Theilmeier G, Michiels C, Raes M, Bult H, Collen D et al (2001) Adenovirus-mediated gene transfer of human platelet-activating factor-acetylhydrolase prevents injury-induced neointima formation and reduces spontaneous atherosclerosis in apolipoprotein E-deficient mice. *Circulation* 103(20): 2495–2500

51 Schober A, Bernhagen J, Thiele M, Zeiffer U, Knarren S, Roller M, Bucala R, Weber C (2004) Stabilization of atherosclerotic plaques by blockade of macrophage migration

inhibitory factor after vascular injury in apolipoprotein E-deficient mice. *Circulation* 109(3): 380–385

52 Schober A, Zernecke A, Liehn EA, von Hundelshausen P, Knarren S, Kuziel WA, Weber C (2004) Crucial role of the CCL2/CCR2 axis in neointimal hyperplasia after arterial injury in hyperlipidemic mice involves early monocyte recruitment and CCL2 presentation on platelets. *Circ Res* 95(11): 1125–1133

53 Shi W, Pei H, Fischer JJ, James JC, Angle JF, Matsumoto AH, Helm GA, Sarembock IJ (2004) Neointimal formation in two apolipoprotein E-deficient mouse strains with different atherosclerosis susceptibility. *J Lipid Res* 45(11): 2008–2014

54 Schulze PC, de Keulenaer GW, Kassik KA, Takahashi T, Chen Z, Simon DI, Lee RT (2003) Biomechanically induced gene iex-1 inhibits vascular smooth muscle cell proliferation and neointima formation. *Circ Res* 93(12): 1210–1217

55 Reis ED, Roque M, Dansky H, Fallon JT, Badimon JJ, Cordon-Cardo C, Shiff SJ, Fisher EA (2000) Sulindac inhibits neointimal formation after arterial injury in wild-type and apolipoprotein E-deficient mice. *Proc Natl Acad Sci USA* 97(23): 12764–12769

56 Kumar A, Lindne V (1997), Remodeling with neointima formation in the mouse carotid artery after cessation of blood flow. *Arterioscler Thromb Vasc Biol* 17(10): 2238–2244

57 Berguer R, Higgins RF, Reddy DJ (1980) Intimal hyperplasia. An experimental study. *Arch Surg* 115(3): 332–335

58 Dobrin PB, Littooy FN, Endean ED (1989) Mechanical factors predisposing to intimal hyperplasia and medial thickening in autogenous vein grafts. *Surgery* 105(3): 393–400

59 Kraiss LW, Kirkman TR, Kohler TR, Zierler B, Clowes AW (1991) Shear stress regulates smooth muscle proliferation and neointimal thickening in porous polytetrafluoroethylene grafts. *Arterioscler Thromb* 11(6): 1844–1852

60 Kohler TR, Kirkman TR, Kraiss LW, Zierler BK, Clowes AW (1991) Increased blood flow inhibits neointimal hyperplasia in endothelialized vascular grafts. *Circ Res* 69(6): 1557–1565

61 Rectenwald JE, Moldawer LL, Huber TS, Seeger JM, Ozaki CK (2000) Direct evidence for cytokine involvement in neointimal hyperplasia. *Circulation* 102(14): 1697–1702

62 Korshunov VA, Berk BC (2003) Flow-induced vascular remodeling in the mouse: a model for carotid intima-media thickening. *Arterioscler Thromb Vasc Biol* 23(12): 2185–2191

63 Sullivan CJ, Hoying JB (2002) Flow-dependent remodeling in the carotid artery of fibroblast growth factor-2 knockout mice. *Arterioscler Thromb Vasc Biol* 22(7): 1100–1105

64 Dimayuga P, Zhu J, Oguchi S, Chyu KY, Xu XO, Yano J, Shah PK, Nilsson J, Cercek B (1999) Reconstituted HDL containing human apolipoprotein A-1 reduces VCAM-1 expression and neointima formation following periadventitial cuff-induced carotid injury in apoE null mice. *Biochem Biophys Res Commun* 264(2): 465–468

65 von der Thusen JH, van Berkel TJ, Biessen EA (2001) Induction of rapid atherogenesis by perivascular carotid collar placement in apolipoprotein E-deficient and low-density lipoprotein receptor-deficient mice. *Circulation* 103(8): 1164–1170

66 Wu L, Iwai M, Nakagami H, Li Z, Chen R, Suzuki J, Akishita M, de Gasparo M, Horiuchi M (2001) Roles of angiotensin II type 2 receptor stimulation associated with selective angiotensin II type 1 receptor blockade with valsartan in the improvement of inflammation-induced vascular injury. *Circulation* 104(22): 2716–2721

67 Liu HW, Iwai M, Takeda-Matsubara Y, Wu L, Li JM, Okumura M, Cui TX, Horiuchi M (2002) Effect of estrogen and AT1 receptor blocker on neointima formation. *Hypertension* 40(4): 451–457; discussion 448–450

68 Vink A, Schoneveld AH, van der Meer JJ, van Middelaar BJ, Sluijter JP, Smeets MB, Quax PH, Lim SK, Borst C, Pasterkamp G, de Kleijn DP (2002) *In vivo* evidence for a role of toll-like receptor 4 in the development of intimal lesions. *Circulation* 106(15): 1985–1990

69 Chyu KY, Dimayuga P, Zhu J, Nilsson J, Kaul S, Shah PK, Cercek B (1999) Decreased neointimal thickening after arterial wall injury in inducible nitric oxide synthase knock-out mice. *Circ Res* 85(12): 1192–1198

70 Moroi M, Izumida T, Morita T, Tatebe J, Ishii C, Imai T, Yagi S, Yamaguchi T, Katayama S (2003) Effect of p53 deficiency on external vascular cuff-induced neointima formation. *Circ J* 67(2): 149–153

71 von der Thusen JH, van Vlijmen BJ, Hoeben RC, Kockx MM, Havekes LM, van Berkel TJ, Biessen EA (2002) Induction of atherosclerotic plaque rupture in apolipoprotein E$^{-/-}$ mice after adenovirus-mediated transfer of p53. *Circulation* 105(17): 2064–2070

72 Von Der Thusen JH, Kuiper J, Fekkes ML, De Vos P, Van Berkel TJ, Biessen EA (2001) Attenuation of atherogenesis by systemic and local adenovirus-mediated gene transfer of interleukin-10 in LDLr$^{-/-}$ mice. *FASEB J* 15(14): 2730–2732

73 Patel SS, Thiagarajan R, Willerson JT, Yeh ET (1998) Inhibition of alpha4 integrin and ICAM-1 markedly attenuate macrophage homing to atherosclerotic plaques in ApoE-deficient mice. *Circulation* 97(1): 75–81

74 Nakashima Y, Plump AS, Raines EW, Breslow JL, Ross R (1994) ApoE-deficient mice develop lesions of all phases of atherosclerosis throughout the arterial tree. *Arterioscler Thromb* 14(1): 133–140

75 Aicher A, Heeschen C, Mohaupt M, Cooke JP, Zeiher AM, Dimmeler S (2003) Nicotine strongly activates dendritic cell-mediated adaptive immunity: potential role for progression of atherosclerotic lesions. *Circulation* 107(4): 604–611

76 Kim CJ, Khoo JC, Gillotte-Taylor K, Li A, Palinski W, Glass CK, Steinberg D (2000) Polymerase chain reaction-based method for quantifying recruitment of monocytes to mouse atherosclerotic lesions *in vivo*: enhancement by tumor necrosis factor-alpha and interleukin-1 beta. *Arterioscler Thromb Vasc Biol* 20(8): 1976–1982

77 Van Lenten BJ, Wagner AC, Anantharamaiah GM, Garber DW, Fishbein MC, Adhikary L, Nayak DP, Hama S, Navab M, Fogelman AM (2002) Influenza infection promotes macrophage traffic into arteries of mice that is prevented by D-4F, an apolipoprotein A-I mimetic peptide. *Circulation* 106(9): 1127–1132

78 Navab M, Anantharamaiah GM, Reddy ST, Van Lenten BJ, Hough G, Wagner A, Naka-

mura K, Garber DW, Datta G, Segrest JP (2003) Human apolipoprotein AI mimetic peptides for the treatment of atherosclerosis. *Curr Opin Investig Drugs* 4(9): 1100–1104

79 Madjid M, Naghavi M, Malik BA, Litovsky S, Willerson JT, Casscells W (2002) Thermal detection of vulnerable plaque. *Am J Cardiol* 90(10C): 36L–39L

80 Verheye S, De Meyer GR, Van Langenhove G, Knaapen MW, Kockx MM (2002) In vivo temperature heterogeneity of atherosclerotic plaques is determined by plaque composition. *Circulation* 105(13): 1596–1601

81 Pasterkamp G, Falk E (2000) Atherosclerotic plaque rupture: An overview. *J Clin Basic Cardiol* 3(2): 81–86

82 Rosenfeld ME, Polinsky P, Virmani R, Kauser K, Rubanyi G, Schwartz SM (2000) Advanced atherosclerotic lesions in the innominate artery of the ApoE knockout mouse. *Arterioscler Thromb Vasc Biol* 20(12): 2587–2592

83 Johnson JL, Jackson CL (2001) Atherosclerotic plaque rupture in the apolipoprotein E knockout mouse. *Atherosclerosis* 154(2): 399–406

84 Rekhter MD, Hicks GW, Brammer DW, Work CW, Kim JS, Gordon D, Keiser JA, Ryan MJ (1998) Animal model that mimics atherosclerotic plaque rupture. *Circ Res* 83(7): 705–713

85 Rekhter MD, Hicks GW, Brammer DW, Hallak H, Kindt E, Chen J, Rosebury WS, Anderson MK, Kuipers PJ, Ryan MJ (2000) Hypercholesterolemia causes mechanical weakening of rabbit atheroma: local collagen loss as a prerequisite of plaque rupture. *Circ Res* 86(1): 101–108

86 Rekhter M (2002) Vulnerable atherosclerotic plaque: emerging challenge for animal models. *Curr Opin Cardiol* 17(6): 626–632

87 Biasucci LM, Liuzzo G, Fantuzzi G, Caligiuri G, Rebuzzi AG, Ginnetti F, Dinarello CA, Maseri A (1999) Increasing levels of interleukin (IL)-1Ra and IL-6 during the first 2 days of hospitalization in unstable angina are associated with increased risk of in-hospital coronary events. *Circulation* 99(16): 2079–2084

88 Ridker PM, Rifai N, Pfeffer M, Sacks F, Lepage S, Braunwald E (2000) Elevation of tumor necrosis factor-alpha and increased risk of recurrent coronary events after myocardial infarction. *Circulation* 101(18): 2149–2153

89 Biasucci LM, Vitelli A, Liuzzo G, Altamura S, Caligiuri G, Monaco C, Rebuzzi AG, Ciliberto G, Maseri A (1996) Elevated levels of interleukin-6 in unstable angina. *Circulation* 94(5): 874–877

90 Liuzzo G, Biasucci LM, Gallimore JR, Grillo RL, Rebuzzi AG, Pepys MB, Maseri A (1994) The prognostic value of C-reactive protein and serum amyloid a protein in severe unstable angina. *N Engl J Med* 331(7): 417–424

91 Andrassy M, Belov D, Harja E, Zou YS, Leitges M, Katus HA, Nawroth PP, Yan SD, Schmidt AM, Yan SF (2005) Central role of PKCbeta in neointimal expansion triggered by acute arterial injury. *Circ Res* 96(4): 476–483

92 Hutter R, Sauter BV, Reis ED, Roque M, Vorchheimer D, Carrick FE, Fallon JT, Fuster V, Badimon JJ (2003) Decreased reendothelialization and increased neointima formation

with endostatin overexpression in a mouse model of arterial injury. *Circulation* 107(12): 1658–1663

93   Hofmann CS, Sullivan CP, Jiang HY, Stone PJ, Toselli P, Reis ED, Chereshnev I, Schreiber BM, Sonenshein GE (2004) B-Myb represses vascular smooth muscle cell collagen gene expression and inhibits neointima formation after arterial injury. *Arterioscler Thromb Vasc Biol* 24(9): 1608–1613

94   Zhu B, Reardon CA, Getz GS, Hui DY (2002) Both apolipoprotein E and immune deficiency exacerbate neointimal hyperplasia after vascular injury in mice. *Arterioscler Thromb Vasc Biol* 22(3): 450–455

95   De Geest B, Zhao Z, Collen D, Holvoet P (1997) Effects of adenovirus-mediated human apo A-I gene transfer on neointima formation after endothelial denudation in apo E-deficient mice. *Circulation* 96(12): 4349–4356

96   Chen Z, Fukutomi T, Zago AC, Ehlers R, Detmers PA, Wright SD, Rogers C, Simon DI (2002) Simvastatin reduces neointimal thickening in low-density lipoprotein receptor-deficient mice after experimental angioplasty without changing plasma lipids. *Circulation* 106(1): 20–23

97   Simon DI, Dhen Z, Seifert P, Edelman ER, Ballantyne CM, Rogers C (2000) Decreased neointimal formation in Mac-1(-/-) mice reveals a role for inflammation in vascular repair after angioplasty. *J Clin Invest* 105(3): 293–300

98   Kawasaki T, Dewerchin M, Lijnen HR, Vreys I, Vermylen J, Hoylaerts MF (2001) Mouse carotid artery ligation induces platelet-leukocyte-dependent luminal fibrin, required for neointima development. *Circ Res* 88(2): 159–166

99   Li Y, Minamino T, Tsukamoto O, Yujiri T, Shintani Y, Okada K, Nagamachi Y, Fujita M, Hirata A, Sanada S et al (2005) Ablation of MEK kinase 1 suppresses intimal hyperplasia by impairing smooth muscle cell migration and urokinase plasminogen activator expression in a mouse blood-flow cessation model. *Circulation* 111(13): 1672–1678

100 Jonsson-Rylander AC, Nilsson T, Fritsche-Danielson R, Hammarstrom A, Behrendt M, Andersson JO, Lindgren K, Andersson AK, Wallbrandt P, Rosengren B et al (2005) Role of ADAMTS-1 in atherosclerosis: remodeling of carotid artery, immunohistochemistry, and proteolysis of versican. *Arterioscler Thromb Vasc Biol* 25(1): 180–185

101 Hassan GS, Jasmin JF, Schubert W, Frank PG, Lisanti MP (2004) Caveolin-1 deficiency stimulates neointima formation during vascular injury. *Biochemistry* 43(26): 8312–8321

102 Khatri JJ, Johnson C, Magid R, Lessner SM, Laude KM, Dikalov SI, Harrison DG, Sung HJ, Rong Y, Galis ZS (2004) Vascular oxidant stress enhances progression and angiogenesis of experimental atheroma. *Circulation* 109(4): 520–525

103 Tran PK, Tran-Lundmark K, Soininen R, Tryggvason K, Thyberg J, Hedin U (2004) Increased intimal hyperplasia and smooth muscle cell proliferation in transgenic mice with heparan sulfate-deficient perlecan. *Circ Res* 94(4): 550–558

104 Kuzuya M, Kanda S, Sasaki T, Tamaya-Mori N, Cheng XW, Itoh T, Itohara S, Iguchi A (2003) Deficiency of gelatinase a suppresses smooth muscle cell invasion and development of experimental intimal hyperplasia. *Circulation* 108(11): 1375–1381

105 Singh R, Pan S, Mueske CS, Witt TA, Kleppe LS, Peterson TE, Caplice NM, Simari RD (2003) Tissue factor pathway inhibitor deficiency enhances neointimal proliferation and formation in a murine model of vascular remodelling. *Thromb Haemost* 89(4): 747–751

106 Singh R, Pan S, Mueske CS, Witt T, Kleppe LS, Peterson TE, Slobodova A, Chang JY, Caplice NM, Simari RD (2001) Role for tissue factor pathway in murine model of vascular remodeling. *Circ Res* 89(1): 71–76

107 Qin F, Impeduglia T, Schaffer P, Dardik H (2003) Overexpression of von Willebrand factor is an independent risk factor for pathogenesis of intimal hyperplasia: preliminary studies. *J Vasc Surg* 37(2): 433–439

108 Squadrito F, Deodato B, Bova A, Marini H, Saporito F, Calo M, Giacca M, Minutoli L, Venuti FS, Caputi AP, Altavilla D (2003) Crucial role of nuclear factor-kappaB in neointimal hyperplasia of the mouse carotid artery after interruption of blood flow. *Atherosclerosis* 166(2): 233–242

109 Galis ZS, Johnson C, Godin D, Magid R, Shipley JM, Senior RM, Ivan E (2002) Targeted disruption of the matrix metalloproteinase-9 gene impairs smooth muscle cell migration and geometrical arterial remodeling. *Circ Res* 91(9): 852–859

110 Cho A, Reidy MA (2002) Matrix metalloproteinase-9 is necessary for the regulation of smooth muscle cell replication and migration after arterial injury. *Circ Res* 91(9): 845–851

111 Murakoshi N, Miyauchi T, Kakinuma Y, Ohuchi T, Goto K, Yanagisawa M, Yamaguchi I (2002) Vascular endothelin-B receptor system *in vivo* plays a favorable inhibitory role in vascular remodeling after injury revealed by endothelin-B receptor-knockout mice. *Circulation* 106(15) 1991–1998

112 Kumar A, Hoover JL, Simmons CA, Lindner V, Shebuski RJ (1997) Remodeling and neointimal formation in the carotid artery of normal and P-selectin-deficient mice. *Circulation* 96(12): 4333–4342

113 Bryant SR, Bjercke RJ, Erichsen DA, Rege A, Lindner V (1999) Vascular remodeling in response to altered blood flow is mediated by fibroblast growth factor-2. *Circ Res* 84(3): 323–328

114 Bot I, von der Thusen JH, Donners MM, Lucas A, Fekkes ML, de Jager SC, Kuiper J, Daemen MJ, van Berkel TJ, Heeneman S, Biessen EA (2003) Serine protease inhibitor Serp-1 strongly impairs atherosclerotic lesion formation and induces a stable plaque phenotype in ApoE$^{-/-}$mice. *Circ Res* 93(5): 464–471

115 Oguchi S, Dimayuga P, Zhu J, Chyu KY, Yano J, Shah PK, Nilsson J, Cercek B (2000) Monoclonal antibody against vascular cell adhesion molecule-1 inhibits neointimal formation after periadventitial carotid artery injury in genetically hypercholesterolemic mice. *Arterioscler Thromb Vasc Biol* 20(7): 1729–1736

116 de Nooijer R, von der Thusen JH, Verkleij CJ, Kuiper J, Jukema JW, van der Wall EE, van Berkel JC, Biessen EA (2004) Overexpression of IL-18 decreases intimal collagen content and promotes a vulnerable plaque phenotype in apolipoprotein-E-deficient mice. *Arterioscler Thromb Vasc Biol* 24(12): 2313–2319

117 Lardenoye JH, Delsing DJ, de Vries MR, Deckers MM, Princen HM, Havekes LM, van

Hinsbergh VW, van Bockel JH, Quax PH (2000) Accelerated atherosclerosis by placement of a perivascular cuff and a cholesterol-rich diet in ApoE*3Leiden transgenic mice. *Circ Res* 87(3): 248–253

118 Imai Y, Shindo T, Maemura K, Sata M, Saito Y, Kurihara Y, Akishita M, Osuga J, Ishibashi S, Tobe K et al (2002) Resistance to neointimal hyperplasia and fatty streak formation in mice with adrenomedullin overexpression. *Arterioscler Thromb Vasc Biol* 22(8): 1310–1315

119 Isoda K, Nishikawa K, Kamezawa Y, Yoshida M, Kusuhara M, Moroi M, Tada N, Ohsuzu F (2002) Osteopontin plays an important role in the development of medial thickening and neointimal formation. *Circ Res* 91(1): 77–82

120 Chen X, Li Z, Li J (2002) Anti-inflammatory effect of cerivastatin in vascular injury independent of serum cholesterol and blood pressure lowering effects in mouse model. *Chin J Traumatol* 5(5): 294–298

121 Isoda K, Shiigai M, Ishigami N, Matsuki T, Horai R, Nishikawa K, Kusuhara M, Nishida Y, Iwakura Y, Ohsuzu F (2003) Deficiency of interleukin-1 receptor antagonist promotes neointimal formation after injury. *Circulation* 108(5): 516–518

Index

Stevenson, C.S., Novartis Institutes for Biomedical Research, Horsham, UK / Marshall, L.A., Johnson and Johnson PRD, Spring House, PA, USA / Morgan, D.W., Biogenldec, Cambridge, MA, USA (Eds)

In Vivo Models of Inflammation

2nd Edition
Volume I

2006. 219 p. Hardcover
ISBN 3-7643-7519-1
PIR - Progress in Inflammation Research

Contents:

For orders originating from all over the world except USA and Canada:
Birkhäuser Customer Service
c/o SDC
Haberstrasse 7, D-69126 Heidelberg
Tel.: +49 / 6221 / 345 0
Fax: +49 / 6221 / 345 42 29
e-mail: orders@birkhauser.ch

For orders originating in the USA and Canada:
Birkhäuser
333 Meadowland Parkway
USA-Secaucus
NJ 07094-2491
Fax: +1 201 348 4505
e-mail: orders@birkhauser.com

The PIR-Series
Progress in Inflammation Research

Homepage: http://www.birkhauser.ch

Up-to-date information on the latest developments in the pathology, mechanisms and therapy of inflammatory disease are provided in this monograph series. Areas covered include vascular responses, skin inflammation, pain, neuroinflammation, arthritis cartilage and bone, airways inflammation and asthma, allergy, cytokines and inflammatory mediators, cell signalling, and recent advances in drug therapy. Each volume is edited by acknowledged experts providing succinct overviews on specific topics intended to inform and explain. The series is of interest to academic and industrial biomedical researchers, drug development personnel and rheumatologists, allergists, pathologists, dermatologists and other clinicians requiring regular scientific updates.

Available volumes: